MIMESIS
INTERNATIONAL

PHILOSOPHY
n. 73

THE INVENTION OF HUMANISM AND THE RECONFIGURATION OF THE CONCEPT OF NATURE IN THE 19TH CENTURY

Edited by Raffaele Carbone

MIMESIS
INTERNATIONAL

Book published with funds of research project 'Paradoxical Humanism and Environment', PRIN PNRR 2022, Università degli Studi di Napoli Federico II, Dipartimento di Studi Umanistici

www.mimesisinternational.com
e-mail: info@mimesisinternational.com

Isbn: 9788869775260
Book series: *Philosophy,* n. 73
Series Issn: 3103-3865

P.I. C.F. 02419370305

CONTENTS

Raffaele Carbone*

INTRODUCTION
The Invention of Humanism and the Reconfiguration of the Concept of Nature in the 19th Century

Abstract

This introduction takes stock of the criticisms levelled at the concept of 'humanism' and the Western humanistic tradition, questioning the possibility of conceiving a non-anthropocentric humanism that is therefore capable of engaging in dialogue with environmentalism. With this in mind, the text briefly reconstructs the current debate on humanism and the genealogy of the term in the 19th century, showing how, in that century, Marxism and Darwinism called into question the traditional idea of 'nature' and led humanism itself to confront a new vision of the connection and interaction between human beings and other living beings.

Keywords

Bildung – Niethammer – Renaissance – Darwin – Marx

The essays collected in this volume deal with two controversial, problematic concepts, each of which has a complex genealogy – two concepts that seemingly appear to be unrelated. In the following pages, we offer a brief historical and theoretical overview of the genealogy and uses of the two terms.

Let us begin with the first concept, 'humanism'. If we wish to begin with reflections and debate on humanism in this first quarter of the 21st century, we should note that humanism has been attacked

* Università degli Studi di Napoli Federico II.

from a number of directions in recent decades: behind this concept critics perceive an essentialist vision of human nature, a mythology of progress as the full realisation of the potentialities of human beings, the celebration of an autonomous, self-determining subject, and finally a fundamental orientation of the Western cultural tradition (with all its geocultural limitations).[1] Sophie Guérard de Latour argued that, on the one hand, the humanist ideal, with its precise historical roots in the Italian Renaissance, emphasises the individual's capacity for self-determination and the ability to choose the shape of an existence that embodies universal values; on the other, humanism runs the risk of being (and ultimately becoming) 'a sectarian philosophy as soon as it imposes this vision of things on people who do not subscribe to it'.[2]

This kind of attitude towards humanism – considered as 'une vieille lune bourgeoise'[3] – is very common nowadays. Nevertheless, the questioning of Western humanism goes back as far as (at least) the twentieth century. This is not the place to refer to or discuss the well-known criticisms of Heidegger, Foucault or Althusser, each of whom questions humanism from a specific perspective that would require in-depth analysis; rather, we will limit ourselves to a consideration of one particular author whose criticisms continue to

1 See Anne Phillips, *The Politics of the Human* (Cambridge: Cambridge University Press, 2015), pp. 14–15; Ken Plummer, *Critical Humanism: A Manifesto for the 21st Century* (Cambridge/UK–Medford/MA: Polity Press, 2021), pp. 3–7. Plummer admits to sharing a number of these critiques, but he thinks that, despite such attacks, 'there are also many good reasons to defend and develop' the idea of humanism (Plummer, *Critical Humanism*, p. 7). From this perspective, he stakes a claim for the 'plurality' and 'hybridity' of humanisms (Plummer, *Critical Humanism*, pp. 15–16).

2 Sophie Guérard de Latour, 'L'humanisme, une valeur à partager entre différentes cultures. À quelles conditions?', *L'Observatoire*, 49.1 (2017), pp. 25–8 (p. 26). The author develops a severe critique of the humanist ideal, showing that, while it is ethnically situated, it has the ambition to set itself up as a universal standard against which any human group can be 'compared, evaluated and ranked'. Ibid., p. 27.

3 François Hartog, *Départager l'humanité. Humains, humanismes, inhumains* (Paris: Gallimard, 2024), p. 316.

spread in various forms to this day. We are all familiar with Claude Lévi-Strauss's harsh criticism of a humanism derived from the Renaissance. In 'Jean-Jacques Rousseau, fondateur des sciences de l'homme',[4] he affirms:

> We started by cutting man off from nature and establishing him in an absolute reign. We believed ourselves to have thus erased his most unassailable characteristic: that he is first a living being. Remaining blind to this common property, we gave free rein to all excesses. Never better than after the last four centuries of his history could a Western man understand that, while assuming the right to impose a radical separation of humanity and animality, while granting to one all that he denied the other, he initiated a vicious circle. The one boundary, constantly pushed back, would be used to separate men from other men and to claim – to the profit of ever smaller minorities – the privilege of a humanism, corrupted at birth by taking self-interest as its principle and its notion.[5]

In a 1971 text entitled 'Race and Culture', which can be read today in *Le regard éloigné*, Lévi-Strauss returns to the 'separations' and 'frontiers' created by Western humanism: he reaffirms that the isolation of man from the rest of nature prepares the ground for the drawing of an ideological line that separates the West from the rest of the world.[6] In summary, it is interesting to note that, according to Lévi-Strauss, Western humanism produces two ruptures: it separates man from the rest of nature and draws a border between western cultures and all other cultures, favouring a hierarchisation of human groups.

---

4 It is a speech given in Geneva on 28 June 1962 during the ceremonies marking the 250th anniversary of the philosopher's birth; the text was later included in *Anthropologie structurale deux* (1973).

5 Claude Lévi-Strauss, 'Jean-Jacques Rousseau, Founder of the Sciences of Man', in Id., *Structural Anthropology*, volume 2, trans. by Monique Layton (Chicago: University of Chicago Press, 1983), p. 41. See Lévi-Strauss, 'Jean-Jacques Rousseau, fondateur des sciences de l'homme', *Anthropologie structurale deux* (Paris: Plon, 1996), pp. 45–56 (p. 53).

6 Claude Lévi-Strauss, 'Race et culture', in Id., *Le regard éloigné* (Paris: Plon, 2001), pp. 24–48 (p. 46).

As is well known, starting in the 1990s, Bruno Latour, *mutatis mutandis*, also understood the establishment of humanism as an act of separation. Noting that modernity is often defined as 'humanism' (both to announce the 'birth' of man and to sanction 'his death'), he points out that this modernity/humanism has also led to 'the simultaneous birth of "nonhumanity" – things, or objects, or beasts – and the equally strange beginning of a crossed-out God, relegated to the sidelines'.[7] According to Latour, so long as humanism is conceived in opposition to the object (the nonhuman entities), which has been left to epistemology, 'neither the human nor the nonhuman can be understood'.[8] The problem is the 'absolute distinction' between nature and society, between 'objective nonhumans' and 'free societies', that reject any kind of mediation.[9]

Some scholars, who thematised the concept of 'humanism' in the first quarter of the 21st century, stressed the difficulties the concept raises for them today. In his book *Humanism: Foundations, Diversities, Developments* (2021), Jörn Rüsen states, on the one hand, that humanism tends to highlight the cultural quality of human beings, by which they essentially differ from nature.[10] On the other, he notes that humanism cannot be reduced to 'an ideology of human

7 Bruno Latour, *Nous n'avons jamais été modernes* (Paris: La Découverte/Poche, 1997) p. 23; *We Have Never Been Modern*, trans. by Catherine Porter (Cambridge, MA: Harvard University Press, 1993), p. 13.

8 Latour, *Nous n'avons jamais été modernes*, p. 186; *We Have Never Been Modern*, p. 136.

9 Latour, *Nous n'avons jamais été modernes*, p. 192; *We Have Never Been Modern*, p. 140.

10 'In the modern understanding of humanism, nature is a counter-concept of humanity. By speaking of "humanity" as a principle of human self-understanding and as a guideline of human practical life, humanism emphasizes the cultural quality of men, by which it essentially differs from nature, let alone from all other species in the visible world. Herder expressed this trans-natural "nature" of humankind by addressing men as "Freigelassene der Schöpfung" (released slaves of the creation). This should be understood as being liberated from the constraints of nature, from the submission to the commands of its laws'. Jörn Rüsen, *Humanism: Foundations, Diversities, Developments* (London–New York: Routledge, 2021), Chapter 8: 'Humanism and nature – some reflections on a complex relationship', pp. 91–100 (p. 92).

mastery over nature, which finally ends in the ecological catastrophe of today'.[11] Indeed, according to him, 'humanism is an elaborated and historically deeply rooted form of [a normative] cultural orientation' because the specific character of human action is that it is based on 'purposes, intentions, culturally articulated needs'. Humanism deals with the ability of every human being 'to bring about its competence for this orientation by a special activity'. This activity is perfectly expressed by the German word *Bildung*, which Jörn Rüsen translates not as 'education', but as 'self-cultivation'. 'This "Bildung" means indeed self-empowerment, but an empowerment to receive the power of value-guided activity'.[12] Nevertheless 'Bildung' and culture presuppose our body, which constitutes the insuppressible connection with nature: it is 'the natural place for manifesting culture; it is a necessary condition of bringing it into existence, of coining a human face onto nature'.[13] Finally, according to Jörn Rüsen, the internal tensions in humanism should 'encourage us to look at the human nature as an already achieved synthesis of nature and culture'.[14]

Today – without forgetting the idea of a more democratic form of humanism based on an enhanced dialogue between cultural traditions, which was defended by Edward Said[15] –, there is no shortage of defenders of humanism, and even within the field of literature and environmental struggles there are voices claiming a

---

11 Ibid., p. 93. In its ancient origins and modern developments, 'Western humanism included the idea that man subjugates and dominates nature. This way of thinking reflects the urban civilization of a city, where human life is no longer close to nature though its original base was grounded on agriculture'. Rüsen then evokes Pico della Mirandola: his Oratio is the epitome of a dominant idea in modern times: the idea that 'God has put the humans into the center of the world and has entitled them to relate the whole creation to themselves, since in humankind the cosmos reflects itself in its sense-bearing order. At the same time the humans are entitled to create the order of their world on their own'. Ibid., p. 94.

12 Ibid., p. 96.

13 Ibid., p. 98.

14 Ibid., p. 99.

15 Edward W. Said, *Humanism and Democratic Criticism* (Basingstoke: Palgrave Macmillan, 2004).

humanism in harmony with an environmentalist theory and approach. This kind of humanism calls for action on the environment which should be rooted in the fundamental belief, not of human superiority, but of human responsibility.[16] According to Lewis P. Hinchman, by stressing the critical and historical orientation of the Renaissance humanist tradition, it is possible to 'trace a line of development that runs from Renaissance humanism to both the historical natural sciences and ecology, as well as to the humanities, while distinguishing both from sciences in search of abstract, universal laws'.[17] The study of classical languages, cultures and literature does not have an antiquarian character, but allows us, through the study of the past, to create pathways between worlds that are distant in time, and therefore reflect on their similarities and differences, making it possible to realise a 'fusion of horizons.'[18] Furthermore, consider this presentation of 'romantic humanism':[19]

> Here, as in the Renaissance, humanism protests against abstract universality, stresses the development of persons and cultures toward goals immanent in them yet still inchoate, and comes out in favour of the unique, the local, the historically contingent. That approach would eventually form the ethical background of the environmentalists' commitment to both biodiversity and cultural pluralism. Humanist historicism has bequeathed to us a sensitivity to the uniqueness of species, languages and cultures; to lose any of them means losing an irreplaceable treasure, an entire mode of being that arose spontaneously and can never be recreated by artifice. Herder, the romantic humanist, was the first to articulate the ethical imperative of preserving historical uniqueness.[20]

16 https://heritage.humanists.uk/article/a-brief-history-of-humanist-environmentalism/
17 Lewis P. Hinchman, 'Is Environmentalism a Humanism?', *Environmental Values*, 13.1 (2004), 3–29 (p. 9).
18 Ibid., p. 8.
19 Herbert Arthur Hodges, *Wilhelm Dilthey: An Introduction* (London: Routledge & Kegan Paul, 1949), p. 2. Quoted by Hinchman, 'Is Environmentalism a Humanism?', p. 9.
20 Hinchman, 'Is Environmentalism a Humanism?', p. 11.

Alongside the detractors, among the not too few thinkers who today claim the centrality of the concept of humanism in understanding our time and our past, we would also like to point out Francis Wolff's proposal: his intention to rebuild humanism and universalism on the basis of the anthropological disposition to language, i.e. man's ability 'to reason together [raisonner ensemble]'.[21] 'This universality – Wolff says – goes beyond the universality of culture, because the diversity of languages never prevents dialogue, whereas cultural diversity is often an obstacle to mutual understanding'.[22]

***

As for the second concept of this book, 'nature', it would be impossible even to summarise its history and meanings. At the dawn of modernity, in the 1588 edition of *The Essays*, Montaigne considered 'Nature' akin to 'a gentle guide, but no more gentle than wise and just'[23] and as a 'great and all-powerful Giver'.[24] In another chapter, 'Education of children', Montaigne evokes 'the great picture of our mother Nature in her full majesty [cette grande image de nostre mere nature en son entiere magesté]':[25] here he suggests that we can have access to an imaginary notion of the whole and rise to the idea of nature-totality embracing all things, of which man, his thoughts, his beliefs and his judgements are only but a small part.

In the present-day, the anti-speciesist activist and theorist Yves Bonnardel affirms that nature is 'a notion that sees the world as an ordered, or at least balanced, whole, in which all "natural" elements

21 Francis Wolff, *Plaidoyer pour l'universel. Fonder l'humanisme* (Paris: Librairie Arthème Fayard, 2021), p. 159.

22 'Cette universalité va au-delà de l'universalité de la culture, parce que la diversité des langues n'empêche jamais le dialogue tandis que la diversité culturelle est souvent un obstacle à l'intercompréhension'. Ibid.

23 *The Complete Essays of Montaigne*, translated by Donald Frame (Stanford: Stanford University Press, 1958), chapter III, 13, 'Of experience', p. 855. 'Nature est un doux guide, mais non pas plus doux que prudent et juste'. *Les Essais*, ed. by Pierre Villey and Louis-Verdun Saulnier (Paris: PUF, 2004), III, 13, p. 1113 b.

24 Ibid.

25 Ibid., 'Education of children', p. 116; *Les Essais*, I, 26, p. 157.

have a "natural" place, and thus contribute, by playing their part, to the harmony of the whole. It always serves more or less as a model, a standard'.[26] In summary, even though such a notion of nature is presented as a description, it is a more or less veiled prescription;[27] 'nature is supposed to be everything that exists – apart from the humans who are supposed to denote it. It is the totality, perceived as a power, a kind of vast organism, an order or balance functioning harmoniously. As a totality, "order" is invested with a value infinitely greater than that accorded to each of the "elements" that make it up.'[28] Bonnardel proposes a fundamental rethinking of our current vision of nature: a critique of morality and hierarchical systems, a rejection of the idea of human identity, and a demand for a non-speciesist environmentalism.[29] It is no longer a question of preserving nature, but of putting the interests of other sentient beings at the forefront, just as much as those of humans, in benefiting from an environment that is in good shape, whatever that may be[30].

This book, and the conference that preceded it, stem partly from the insights gained from an exhibition held at the Musée d'Orsay between May and July 2021, entitled *Les origines du monde: L'invention de la nature au XIXe siècle*. In the prologue of the exhibition and of the catalogue, the organisers stated that in order to think about man's relationship with Nature, it is essential to take a detour via the 19th century. During the 19th century, reflections on the relationship between humans and the natural world – which had already undergone a major reconfiguration thanks to Linnaeus' classifications and Buffon's idea of a natural history of man in the 18th century – was radically transformed: the cataloguing of land,

26 Yves Bonnardel, 'Idée de Nature, humanisme et négation de la pensée animale'. Brochure éditée par 'un réseau contre le spécisme', 20 rue Cavenne, 69007 Lyon, 1–23 (p. 2). Here and hereafter, if the English translation is not indicated, the translation is our own.

27 Ibid., pp. 2–3.

28 Ibid., p. 3.

29 Ibid., p. 20.

30 Yves Bonnardel, 'Contre l'apartheid des espèces', in *Espèces et éthique. Darwin: une (r)évolution à venir*, ed. by Yves Bonnardel, David Olivier, James Rachels and Estiva Reus (Lyon: tahin party, 2001).

plants, and animals came to cover all continents; the age of the earth and of life was discovered; landscapes were profoundly changed by industrialisation and urbanisation. Furthermore, the 19th century saw the birth of the modern natural sciences: biology, palaeontology, organic chemistry, physiology, geology, bacteriology, anthropology, and ecology. But it is above all the theories of evolution that make it possible to redefine the place of human beings in nature, reconfigure the links that connect them to the animal world, and rethink their very animality in a world understood as an ecosystem[31]. Starting with *The Origin of Species*, published by Darwin in 1859, human beings are part of a family tree that encompasses all living beings. In 1863, Thomas Huxley, the leading naturalist and populariser of Darwinian theory, published *Evidence as to Man's Place in Nature* and *Six Lectures To Working Men*: according to him, the question of man's place in nature is the fundamental question posed to man.[32]

At the beginning of the second chapter of *Evidence as to Man's Place in Nature*, Thomas Huxley states:

> The question of questions for mankind – the problem which underlies all others, and is more deeply interesting than any other – is the ascertainment of the place which Man occupies in nature and of his relations to the universe of things. Whence our race has come; what are the limits of our power over nature, and of nature's power over us; to what goal we are tending; are the problems which present themselves anew and with undiminished interest to every man born into the world.[33]
>
> For it will be admitted that some knowledge of man's position in the animate world is an indispensable preliminary to the proper understanding of his relations to the universe – and this again resolves itself, in the long run, into an inquiry into the nature and the closeness of

31 See Laura Bossi, 'La nature dévoilée par la science', in *Les origines du monde. L'invention de la nature au XIXe siècle*, ed. by Laura Bossi (Paris: Musée d'Orsay/Gallimard, 2020), pp. 20–3 (p. 21).

32 See Dominique Lestel, *Les origines animales de la culture* (Paris: Flammarion, 2009), p. 28.

33 Thomas H. Huxley, *Evidence as to Man's Place in Nature* (New York: D. Appleton, 1863), p. 71.

> the ties which connect him with those singular creatures whose history has been sketched in the preceding pages.[34]

It would also be interesting to reconstruct the history of the Western vision of nature. The organisers of the Musée d'Orsay exhibition noted that until the 18th century, the world was seen as a garden in which nature served man. During the Renaissance, precise descriptions of plants and animals proliferated and the first illustrated treatises on zoology appeared. Princes, humanists and scholars created 'cabinets of curiosity' combining naturalia (natural objects), artificialia (man-made objects) and mirabilia (astonishing and marvellous objects). Voyages of exploration led to the collection and acclimatisation of species unknown in Europe in princely gardens and menageries. The fascination of large animals (rhinoceroses, giraffes, etc.) inspired artists. The collection, inventory, description and classification of minerals, plants and animals seemed to reiterate Adam's naming of animals and bring the diversity of the 'new worlds' into the orbit of Christian civilisation.[35]

The naming of species was at the heart of the work of the Swedish naturalist Linnaeus (1707–1778), the father of the binomial nomenclature of species (a combination of two Latin names) and of the classification into class, order, genus, species and variety. The Comte de Buffon (1707–1788), Director of the Jardin du Roi and of the Cabinet d'histoire naturelle (1739), also aspired to a complete inventory of Nature, but he abandoned classifications, which he considered arbitrary, and sought the natural causes of phenomena.[36] His *Histoire naturelle* in 36 illustrated volumes (1749–1789) was a huge success.

Of course, in the 19th century a radically new conception of nature and, consequently, of the link between humanity and nature emerged: a conception that called into question the teleological vision and the stylised image of nature as a set of resources and instruments that

---

34 Ibid., p. 73.

35 These themes were tackled in the exhibition at the Musée d'Orsay (19th May-18th July 2021). See the catalogue.

36 See Catherine and Raphaël Larrère, *Du bon usage de la nature. Pour une philosophie de l'environnement* (Paris: Flammarion, 2022), pp. 78–9.

man could use, aimed at his fulfilment and the improvement of his living conditions.[37]

***

On the other hand, it seems significant, if not paradoxical, that in the very century in which the concept of nature was 'invented' or 'reconfigured', the term 'humanism' actually began to be used in pedagogical, philosophical and literary contexts.

Indeed, it is in the second half of the 18th century, in an article by Quesnay that appeared in *Éphémérides du citoyen*,[38] that the word 'humanism' entered the European philosophical-cultural lexicon with the meaning of 'general love of humanity', as a synonym for philanthropy; it is the 19th century, however, that marked a decisive turning point in the concept through, firstly, the publication of the philosopher and educationalist Friedrich Immanuel Niethammer's (1766–1848) book, *The Dispute of Philanthropinism and Humanism in the Theory of Educational Schooling of Our Time*. Niethammer forged the term 'Humanismus' from the Ciceronian *humanitas* and used it as a weapon in a pedagogical battle, pitting the classical studies of the *Gymnasium* against the *Realschule*, which focused on the professional training of the bourgeois. He argued that 'Philanthropinismus'[39] should be called 'Animalismus' since it

37 Consider, for example, Giovanni Pico della Mirandola, Francis Bacon, Descartes, in particular the *Discours de la méthode*, bearing in mind that, as is well known, the culture of the early modern age is more complex than certain simplifications would have us believe: reflect on the critiques of anthropocentrism and teleological views of nature developed by Montaigne, Bruno and Spinoza.

38 François Quesnay, 'Éphémérides du citoyen, ou Bibliothèque raisonnée des sciences morales et politiques', 16.1 (1765), 241–56 (p. 247).

39 As Sebastiaan van Bommel states, 'with the term "philantropinism", Niethammer referred to a powerful current in Enlightenment pedagogy, now commonly known under the name *Philantropismus* (or English "Philanthropinism"), which strongly favoured *real-* above classical education'. Sebastiaan van Bommel, *Classical Humanism and the Challenge of Modernity: Debates on Classical Education in 19th-Century Germany* (Berlin: De Gruyter, 2015), p. 112. Humanistic education

advocates a model of education not on humanity but on animality: it addresses the animal part of man.[40] Niethammer challenged this 'materialist' model and argued that man is not a simple juxtaposition of animality and rationality, but a unified whole from the two elements,[41] opposing 'Humanismus' to 'Philanthropinismus'. According to the humanistic model, the educational process includes an intellectual and moral formation that, inspired by Kant's thought, has man as its goal with a view to advancing mankind. This type of education is to be realised, through the study of Greek and Latin, in the Gymnasium and at university.[42] Indeed, the core of the conflict between philanthropinismus and humanism lay in the debate about the teaching of languages, especially classical languages.[43]

While retaining its place in pedagogical debates, the concept of humanism spread to other contexts as early as the 1830s and 1840s. It was enriched with more complex implications to the extent

---

focuses on 'the life of ideas', on 'liberal knowledge' of 'the true, the good and the beautiful', while 'philantropinist' education focuses chiefly 'on the surrounding material world'. Ibid.

40 Friedrich Immanuel Niethammer, *Der Streit des Philanthropinismus und Humanismus in der Theorie des Erziehungs-Unterrichts unsrer Zeit* (Jena: Frommann, 1808), p. 8.

41 Ibid., p. 67. On this topic, see Stéphane Toussaint, *Humanismes/ Antihumanismes: De Ficin à Heidegger*, Tome I: *Humanitas* et Rentabilité (Paris: Les Belles Lettres, 2008), pp. 73–7.

42 See François Hartog, 'Lettre sur l'Humanisme et les Humanités', *Anabases. Traditions et réceptions de l'Antiquité, Études sur la circulation et la réception des savoirs offertes à Pascal Payen. Actualités et débats*, 35 (2022), 227–43 (p. 232). Those who were later referred to by historiography as 'neo-humanists' soon developed a strong awareness of their belonging to a specific current of thought as well as their filiation with ancient culture, in particular with Greek *paideia*. The term 'neo-humanism' was not actually introduced until 1885, when it was coined by the historian Friedrich Paulsen to distinguish the German educational movement from Renaissance humanism and emphasise its specificity. See Mathilde Lerenard, 'La querelle des philanthropinistes et néohumanistes: pratiques et débats éducatifs en Allemagne autour de 1800', *Essais*, 4 (2014), 81–98.

43 See Kristina Bosakova and Martina F. Bykova, 'Hegel and Niethammer on the Educational Practice in Civil Society', *Journal of Philosophy of Education*, 55.1 (2021), 99–125 (pp. 115–16).

that authors such as Arnold Ruge (1802–1880) reinterpreted it by highlighting a set of social attitudes and positions that it should entail. Along these lines, newspapers also began to attribute new meanings to the concept of humanism in the period before the revolutions of 1848/49.[44]

Also, towards the middle of the 19th century, the term 'humanismus' began to be used to denote a historical event, and an intellectual phenomenon associated with the Renaissance.[45]

Finally, the great scientific and social revolutions of the 19th century (Marx's and Darwin's theories) intersect with the concept and problem of humanism and bring into play previous conceptions of the human being, of his essential characteristics or activities. This was grasped, albeit from a somewhat narrow and simplifying perspective, by the Italian literary critic, scholar and politician Francesco De Sanctis. In his 1883 lecture entitled *Darwinism in art* he claims:

---

44 On these topics see Hans Erich Bödeker, 'Menschheit, Humanismus, Humanität', in *Geschichtliche Grundbegriffe: Historisches Lexikon zur politisch- sozialen Sprache in Deutschland*, ed. by Otto Brunner, Werner Conze, and Reinhart Koselleck (Stuttgart: Klett-Cotta, 1982), vol. 3, pp. 1063–1128; Georg Bollenbeck, *Bildung und Kultur: Glanz und Elend eines deutschen Deutungsmusters* (Frankfurt am Main: Insel, 1994), pp. 142–55; Heidi Hakkarainen, 'Contagious Humanism in Early Nineteenth-Century German-Language Press', *Contributions to the History of Concepts*, 15.1 (2020), 22–46 (pp. 23–4).

45 Karl Hagen, *Deutschlands literarische und religiöse Verhältnisse im Reformationszeitalter. Mit besonderer Rücksicht auf Wilibald Pirkheimer*, 3 vols (Erlangen: Palm, 1841–1844), Vol. 1, p. 79 and *passim*; Georg Voigt, *Die Wiederbelebung des classischen Alterthums oder das erste Jahrhundert des Humanismus*, 2 vols (Berlin: Reimer 1859); Jacob Burckhardt, *Die Kultur der Renaissance in Italien. Ein Versuch*, chapter III, 4: 'Der Humanismus im 14. Jahrhundert', in Id., *Das Geschichtswerk* (Frankfurt am Main: Zweitausendeins, s.d.), Vol. 1, p. 476–80. See Hendrik Birus, 'The Archeology of "humanism"', *Surfaces*, 4 (1994), https://doi.org/10.7202/1064969ar; Mario Todte, *Georg Voigt (1827–1891): Pionier der historischen Humanismusforschung* (Leipzig: Leipziger Universitätsverlag, 2004); Rémi Brague, *Le propre de l'homme* (Paris: Flammarion, 2015).

> We had humanism: today we have an exaggerated animalism. It is clear that in this new environment there is something low and corrupt that wants to be purified. And this will be, if our spirit is willing to look at man less in his already absorbed similarities, and more in his differences that give him the right to say: I am a man and not an animal[46].

In light of this brief overview of the categories of 'humanism' and 'nature', we believe that, in order to understand better what is at stake in current debates on humanism and the claims of a new humanism, also taking the criticism addressed thereto into consideration, it is useful to reconstruct and rethink a genealogy of the concept of humanism. From this perspective, the period between the end of the 18th century and the whole of the 19th century is crucial: in this period, figures such as Marx, Darwin or Nietzsche, for example, produced theoretical and practical revolutions that led Western culture to radically rethink the human and, perhaps, the very concept of humanism that was born at the beginning of the century. The articles collected in this volume aim to take stock of a number of significant authors and moments in the discussion on humanism and the concept of nature during the period under consideration; they also shed light on this crucial era for the interconnections between the concept of 'humanism' and that of 'nature'. The articles can be divided into four areas: the texts in the first section address the question of the genealogy of humanism and the specific approaches of humanism in particular contexts between the 18th and 19th centuries (Stengel, Moreira, Ruoppo); those in the second section focus on the relationship between man and nature, between humanism and naturalism in Marx and Marxism (Petrucciani, Viparelli); those in the third area address the theme of humanism in the light of Darwin's theory of evolution (Bachofen, Abbadessa, Visone, Carbone, Mascolo); those in the last group investigate possible developments in thinking about humanism and nature in a number of significant authors of the 19th and early 20th centuries (Specq, Bocchetti, Carrano).

46 Francesco De Sanctis, *Il darwinismo nell'arte* (Naples: Stabilimento Tipografico dei Classici Italiani, 1883), p. 18.

Friedemann Stengel*

WHAT IS HUMANISM?
Historical and Critical Perspectives[1]

Abstract

'Humanism' is among the most powerful terms in historical and contemporary political, religious, and philosophical debates. 'Humanism' is equally used historiographically and normatively. The term serves to position in ideological conflicts and to cement a claim to interpretation in global historical discourses, but is highly contradictory. Since the middle of the 19th century, a contested field of tension has been opened up between 'religion' and 'humanism' in particular, which was – and is – just as effective politically as it has fundamentally influenced the understanding of religion. My paper addresses 'humanism' in its complexity and striking contradictions. Contemporary definitions of the term are confronted with the 15th to late 19th century historical contexts the term 'humanism' is applied to. Based on Friedrich Immanuel Niethammer's invention of 'humanism' as a pedagogical-philosophical and, according to Niethammer's self-understanding, anti-Enlightenment concept (1808), I will not present an outline of a mere conceptual history of 'humanism', but rather a theoretically oriented discourse, an examination of the contextual front positions, between which 'humanism' has been constructed as a reformist educational concept, a socialist and atheist programme (Marx/Engels, Ruge), a social movement, a confessionalist slogan and an equally anti-reformist and anti-Roman category. These

* Martin-Luther-Universität Halle-Wittenberg.

1 This essay is based on Friedemann Stengel, *Genealogie des Humanismus, Debatten – Kritik – Neue Perspektiven* (Bielefeld: Transcript, 2024), will be published in the Cambridge Elements 2026 series (Elements in the History of Philosophy and Theology in the West) as: *A Critical Genealogy of Humanism* (DOI: 10.1017/9781009634236).

conceptualizations from Niethammer to the German historiography in the middle (Karl Hagen) and at the end of the 19th century (Georg Voigt) have been unmistakably steered and shaped by their contexts. This way, the historiographical – and heuristic – impossibility of the term is shown, which, however, is rooted more precisely in its strict contextuality. In a second part, a) historiographical alternatives to this dilemma are pointed out regarding the use of the term for the 15th and 16th century movement later called 'humanism', in order to finally b) give suggestions not only for an ethical-normative work of historians of 'humanism', but for dealing with 'humanism' in general as a present and future discourse, in connection with discourse-theoretical suggestions and as a decisive question of how man is and how he should be.

Keywords

Humanism – Discourse Theory – Reformation – Hermeticism – Theory of Historiography

1. *Humanism between Historicity and Normativity*

The term 'humanism' appears in academic and media debates across the globe in a variety of languages and cultures. But the inconsistent, polemical, and conflicting ways in which humanism has been defined throughout history persist. Adding to these challenges, there are unresolvable disparities between various definitions of humanisms as they pertain to the Renaissance, the Christian humanisms of the 19th and 20th centuries, the Marxist-Leninist religious policy of the communist party in the former GDR, the Socialist Unity Party of Germany, directed against the churches, which sought to legitimize the categories of reconciliation or – reversed – struggle,[2] or some more recent humanistic associations

2 The ecclesiastical genesis of the concept of humanism in the GDR needs more careful study. Cf. meanwhile Florian Baab, *Was ist Humanismus? Geschichte des Begriffes, Gegenkonzepte, säkulare Humanismen heute*

that portray humanism as an anti-Christian or an exclusively anti-religious worldview.[3] Additionally, the official Russian position, represented by Vladimir Putin, and the official Russian Orthodox position, represented by Patriarch Kirill I, further views humanism as a Western, liberal, heretical ideology linked to homosexuality, capitalism, and anti-Russian tendencies, which must be fought both politically and militarily at all costs. They have been openly propagating this for more than 10 years.[4]

Postcolonial critics such as Dipesh Chakrabarty and Homi Bhabha would point to a seemingly universal humanism that is fundamentally Western-particular and only claims universality.[5] In contrast, one should mention the Christian Zambian Humanism of long-time Zambian President Kenneth Kaunda who has engaged with the

(Regensburg: Pustet, 2013), pp. 129–32; Friedemann Stengel, 'Die SED und das christliche nationale Erbe', *Händel-Jahrbuch*, 59 (2013), pp. 351–9; Nora Blume, 'Humanismus im Kalten Krieg: Behauptung, Zurückweisung und Vereinnahmung von "Humanismus" zwischen Ost und West, Kirche und Staat auf einer Tagung 1959 in Wittenberg', *KZG*, 34.1 (2021), pp. 128–46. Since the 1960s, the SED used 'humanism' either to force the churches to submit to the state or to concede them 'humanistic' goals, in order to present themselves as tolerant. The two strategies were often deliberately interchanged.

3 Andreas Fincke, *Freidenker – Freigeister – Freireligiöse: Kirchenkritische Organisationen in Deutschland seit 1989* (Berlin: EZW, 2002); Id., *Woran glaubt, wer nicht glaubt? Lebens- und Weltbilder von Freidenkern, Konfessionslosen und Atheisten in Selbstaussagen* (Berlin: EZW, 2004).

4 Cf. for the time Olga Tomyuk, 'Zur diskursiven Konstruktion von Humanismus in der Russisch-Orthodoxen Kirche', *KZG*, 36.1 (2023), pp. 138–67.

5 Cf. Dipesh Chakrabarty, 'Humanism in an Age of Globalization', in: *Alltag, Erfahrung, Eigensinn: Historisch-anthropologische Erkundungen*, ed. by Belinda Davis, Thomas Lindenberger and Michael Wildt (Frankfurt/Main, New York: Campus, 2008), pp. 74–90; Homi K. Bhabha, '"Rasse", Zeit und die Revision der Moderne', in: Homi K. Bhabha, *Die Verortung der Kultur* (Tübingen: Stauffenburg, 2000), pp. 353–84 (p. 376); Id., 'Die Frage der Identität: Frantz Fanon und das postkoloniale Privileg', in: ibid., pp. 59–96 (pp. 90–1).

humanism debate of the East German Christian Democratic Union while simultaneously distancing himself from GDR socialism.[6]

We could significantly expand the list of such concepts of humanism, leading to the conclusion that the various understandings not only stand side by side. They are often explicitly opposed to one another and mutually exclusive. We could anticipate the finding that there is no unified, generally accepted concept of humanism neither in the present nor in the past. But this also reveals a dilemma, which I do not wish to remain with: the dilemma that a historical term, which arose in a very specific context, simultaneously contains three dimensions: firstly, a historiographical category that is associated with the claim to express a super-historical current or movement; secondly, a normative category that claims to historically validate or save contemporary normative positions; and thereby thirdly, inscribes current normative positions into the historical term and concept.

I will in a first step illustrate the irreconcilable aporias of the term with a brief overview of the genealogy of humanism. After that, since humanism is often applied from a Western and Central European perspective as a movement from the Renaissance and frequently in conflict with the Reformation, I will propose an alternative approach to this historical concept. Finally, I will suggest how humanism can be made fruitful not as a category, but as a discourse and principle of contention.

2. *Humanism – a Concept against Religion?*

A popular position in Germany, Michael Schmidt-Salomon's so-called evolutionary humanism, is by no means the only one who understands his evolutionary humanism as the culmination of an Enlightenment that has stalled halfway, which can only be realized as a secularization program because religion is seen as a primate

---

6 Cf. Nora Blume, *Humanismus zwischen den Fronten des Kalten Krieges: DDR, Christentum und Befreiungsbewegung in Afrika* (Diploma thesis, University of Halle, 2022), pp. 45–92; Blume, 2021, p. 146.

brain construction.[7] This motif can also be found in the humanism of the SED: Humanism was, by the way, only identified with science, atheism, and above all 'Enlightenment' since the late 1950s. This simplistic explanation has been deeply engrained in the population's consciousness for generations.[8] Even today, a large majority of the East German and certainly also the Central and Eastern European population would answer the question: Are you 'religious' with: No, I am 'humanistic'.

However, a glance at the origins of humanism is striking. Humanism was actually conceived as a counter-program to Enlightenment. For Friedrich Schelling in 1809, 'modern humanism' is not opposed to Christianity and religion but rather stands in opposition to the 'barbaric Enlightenment'.[9] Schelling shared a view of Enlightenment as age of mechanization, utilization, and reduction of humanity to its corporeality, as evidenced in his supposed cultural pessimism was included in a commentary on a text that is considered a significant pedagogical reform program at the start of the 19th century.

3. *The 'Founder' of Humanism: Niethammer*

In the late 18th century there are a few mentions of the term humanism, but it should be distinguished from humanity (humanitas). With his *Streit des Philanthropinismus und Humanismus in der Theorie des Erziehungs-Unterrichts unsrer Zeit* [Dispute between Philanthropinism and Humanism in the Theory of Education of Our Time], Friedrich Immanuel Niethammer is considered the inventor of humanism in 1808.[10] It is a program with clear fronts: a) the 'spirit of

7 Michael Schmidt-Salomon, *Manifest des evolutionären Humanismus: Plädoyer für eine zeitgemäße Leitkultur*, 2nd edn (Aschaffenburg: Alibri, 2006); Baab, pp. 189–217 *passim*.

8 Cf. Stengel, Die SED und das christliche nationale Erbe.

9 Friedrich W. J. Schelling, [Review of:] 'Der Streit des Philanthropinismus und Humanismus in der Theorie des Erziehungsunterrichs unserer Zeit dargestellt von F. I. Niethammer [1808]', in: *Schellings Werke*, 3rd Ergänzungsband (München: Beck, 1968), pp. 457–80 (p. 480).

10 Cf. Stengel, *Genealogie*, pp. 11–24.

humanism'[11] is older than the Enlightenment, and b) for Niethammer, the Enlightenment propagated a worldview that is mechanistic and utilitarian and reduced humanity to function and body, to a 'machine',[12] as seen in the work of Julien Offray de La Mettrie who is not mentioned,[13] but instead of him Frederick II, the Prussian king who promoted that kind of enlightened anthropology which was in Niethammer's eyes represented in the reform pedagogical school *Philanthropinum* in Dessau.[14] Niethammer, who is considered by some to represent a re-Kantianization of idealist philosophy,[15] emphasizes c) the 'dual nature'[16] of man; his 'animality'[17] stands in contrast to humanity as spirituality. This free spirituality is for Niethammer d) not simply intelligibility but it stands against the

11 Friedrich Immanuel Niethammer, *Der Streit des Philanthropinismus und Humanismus in der Theorie des Erziehungs-Unterrichts unsrer Zeit* (Jena: Frommann, 1808), pp. 33–4.

12 Niethammer, p. 43.

13 With his *L'homme machine*, La Mettrie was present in the contemporary discourse on Enlightenment. Cf. Friedemann Stengel, 'Seele, Unsterblichkeit, Auferstehung: Luther im Aufklärungsdiskurs', in: *Das Bild der Reformation in der Aufklärung*, ed. by Wolf-Friedrich Schäufele and Christoph Strohm (Gütersloh: Gütersloher Verlagshaus, 2017), pp. 98–130.

14 Niethammer, p. 15. La Mettrie was a refugee in exile at Frederick's court in Potsdam. Niethammer's term Philanthropinism, instead of Philanthropism, is derived from the reform pedagogical school *Philanthropinum* in Dessau (1774–1793). On Philanthropinum, cf. Christian Eger, *Gute Poesie: Literatur, Kultur und Herrschaft in Dessau-Wörlitz um 1800*, 2 vols (Göttingen: Wallstein, 2025), vol. 1, pp. 535–610, 561–568, 594–597; Michael Rocher, '"Worthmann muss entweder am Gehör oder Verstand oder an beiden einen Fehler haben…": Realitäten des Religionsunterrichts am Dessauer Philanthropin', *BThZ*, 40 (2023), pp. 205–20; Michael Rocher, '"Mit Neuem Eifer an der Bildung Junger Leute Zu Arbeiten": Das Pädagogium Regium Halle und das Philanthropin Dessau im bildungsräumlichen Vergleich' (Halle: Franckesche Stiftungen, 2025).

15 Manfred Frank, *'Unendliche Annäherung': Die Anfänge der philosophischen Frühromantik* (Frankfurt/Main: Suhrkamp, 1997), p. 502.

16 Niethammer, p. 58.

17 Niethammer, p. 34.

de-spiritualization [*Entgeistung*][18] of the Enlightenment project. It points to another reality: life after death and the immortality of man; significantly, the Kantian Niethammer does not use the term 'soul' in this context.[19] In contrast, for Niethammer the Enlightenment fostered the 'refusal to believe in the invisible'[20] and mocked the invisible as 'mystical nonsense belief'.[21] e) Humanism is neither an opposition nor an alternative to Christianity. Humanism is not a spirit of antiquity that was revived during the Renaissance to enter into a relationship with the Reformation. Christianity and humanism are neither a diastasis nor a synthesis; that theme does not exist. Ultimately, f) it concerns the 'education of humankind',[22] with cosmopolitanism being crucial for the Kantian Niethammer. His humanism is not just a linguistic educational program as in later concepts of humanistic grammar schools or gymnasiums. He opposes 'mere *verbiage* [Wortkrams] in pedagogy' and the reduction of philology to the study of '*words and letters*'.[23] Language education

18 Cf. Niethammer, p. 18.

19 Niethammer may have distanced himself from a central theme of Enlightenment theology and philosophy in reaction to Schleiermacher's assertion made only a few years prior that the idea that a person's personality endured after death is wholly irreligious. Niethammer never explicitly affirmed that the soul is immortal even though he spoke of man's awareness of a non-earthly reality and his conviction in a greater purpose for the world and a personal life beyond death. I, therefore, address the question of otherworldliness, afterlife, or supernatural destiny as they pertain to this context in the following. Cf. on the contextual backdrop to this repudiation, Friedemann Stengel, *Enlightenment All the Way to Heaven: Emanuel Swedenborg in the Context of Eighteenth-Century Theology and Philosophy* (West Chester, PA: Swedenborg Foundation, 2023), pp. 826–9; Friedemann Stengel, 'Swedenborg in German Theology in the 1770's and 1780's', in: *Emanuel Swedenborg – Exploring a 'World Memory': Context, Content, Contribution,* ed. by Karl Grandin (Stockholm: Royal Swedish Academy of Sciences, 2013), pp. 334–55 (Contributions to the History of the Royal Swedish Academy of Sciences, vol. 43).

20 Niethammer, p. 48.

21 Niethammer, p. 18.

22 Niethammer, p. 333.

23 Niethammer, p. 166, 164.

is only a 'means to free education'.[24] Niethammer's utopian school of thought focused on the education of reason and the education of man, the enhancement of the human intellect through 'education' and not through mere 'vocational training'.[25] The Greco-Roman antiquity does not play a significant role in Niethammer's writings as he did not simply associate humanism with antiquity, and to that extent, humanism would not have to be revived or recreated parallel to the Renaissance in juxtaposition with the Reformation. Niethammer makes no mention of the Renaissance or the 15th and the 16th centuries.

Thus, Niethammer's humanism serves as a counter-concept against an immanent, functional and utilitarian education in the Enlightenment, against which he emphasized as a normative aspect the intelligible, non-discursive side of humankind. Humanism is conceived as spiritualization and, in this sense, as universal progress. It is not an alternative to religion but directed towards human immortality – which coincidentally parallels the emerging contemporary spiritism around 1800, for example with the famous *Theorie der Geisterkunde* [Theory of Spirit Science, does not mean Pneumatology] by Johann Heinrich Jung-Stilling from 1808.[26] By the way, Niethammer himself reported years later in a letter to Hegel that he was certain of the spiritual presence of his deceased wife through profound emotional experiences, which he regarded as evidence for immortality and a post-mortem reunion.[27] Interestingly, in nearly all texts about Niethammer, the keyword immortality remains unmentioned. I consciously refer to Niethammer because precisely in case of the founder and most important propagator of

---

24 Niethammer, p. 221.

25 Niethammer, p. 188.

26 Johann Heinrich Jung-Stilling, *Theorie der Geister-Kunde, in einer Natur- Vernunft- und Bibelmäsigen Beantwortung der Frage: Was von Ahnungen, Gesichten und Geistererscheinungen geglaubt und nicht geglaubt werden müße* (Nürnberg: Raw, 1808) [reprint Leipzig 1987].

27 Niethammer's letter to Hegel dated 14.07.1832, Gerhard Lindner, *Friedrich Immanuel Niethammer als Christ und Theologe: Seine Entwicklung vom deutschen Idealismus zum konfessionellen Luthertum*, 2 vols (Nürnberg: Verein für bayerische Kirchengeschichte, 1971), p. 291.

humanism collisions with modern and present concepts become visible.

4. *Was the Founder of Humanism a Humanist?*

But what is 'humanism' now? Let us look into the *Enzyklopädie der Neuzeit* [Encyclopedia of Modern Times], one of the newest and perhaps last printed relevant encyclopedias in Germany. 1. It acknowledges that humanism was first anthropologically charged by Niethammer. 2. The 'ideological' identification with the Renaissance is deemed inaccurate, as the so-called *studia humanitatis* do not possess their own anthropological agenda but are merely a canon of subjects. 3. However, the Encyclopedia of Modern Times still claims that humanism was the 'most important European educational movement' of modern times before the Enlightenment.[28] Since it is established in the encyclopedia that 4. humanism has a natural distance from Christianity and aimed at a secular ethics, Niethammer must be excluded from the circle of humanists, and not only Niethammer but also all also figures traditionally considered humanists from the Early Modern period between Erasmus, Reuchlin, and Ficino, as well as many authors from the 19th century or contemporary Christian humanisms. What then should 'humanism' be? A secular entity that 'elegantly' ignored the churches? Christianity-affine or a counter-concept? An 'indicator of an epochal change in Western culture',[29] according to the Encyclopedia of Modern Times? An anthropology or a pedagogical concept arising from a specific critique of Enlightenment that advanced into an anti-church alternative religion in left Hegelianism and later in East Germany? A Marxist-Leninist battle term or a Western hegemonic paradigm? Or an anti-Russian, liberal, LGBTQ ideology?

This confusion of terms is found everywhere. It has arisen from a) the continued use of an intellectual historical construction from the 19th century that b) aims to define a strictly historical concept of

28 Gerrit Walther, Art. 'Humanismus', *Enzyklopädie der Neuzeit*, 5 (2007), pp. 665–92 (pp. 665–6).

29 Walther, p. 691.

humanism in such a way that it c) can also be recognized as a trans- or superhistorical phenomenon.

How can one claim according the *Encyclopedia of Modern Times* that 'humanism' is nevertheless an 'indispensable heuristic tool',[30] when heuristics are based on premises that are historically unverifiable and consist instead of contemporary positions? For it is undisputed that even through the merely heuristic use of terms, fields of investigation are pre-structured.[31] What consequences should be drawn from this observation for today's debate? First, let us take a look at the attributions to humanism after its invention – as epoch, as education program, as 'spirit'.

5. *An Unsolvable Dilemma: Humanism's Life After Its Invention*

5.1. *Humanism before Niethammer?*

In the 18th century, a particular conception of man, which is said to originate from classical antiquity and was invented in the 15th and 16th centuries, is not associated with humanism as a historical and epochal category.[32] The *Zedler*, a very widespread huge encyclopedia from the first decades of 18th century, knows *humaniora* and *studia humanitatis* in the sense of the *artes liberales*, and Zedler understood humanity, not humanism, as 'politeness and affability' [Höflichkeit und Leutseligkeit].[33] Similarly, Kant also moves in this

30 Walther, p. 668.

31 Using Hinduism as an example, it was shown how the conceptualization of a cultural phenomenon pre-structures the empirical phenomena to be investigated, Timothy Fitzgerald, *The Ideology* of *Religious Studies* (New York: Oxford University Press, 2000), p. 19.

32 There is mention of 'humanism' in connection with morality and religion as a 'highly humane' practice in the judiciary in: Johann Wolfgang Goethe. *Aus meinem Leben: Dichtung und Wahrheit*, 3 vols (Tübingen: Cotta, 1811–4), vol. 3, p. 290. Another mention 1798 in Friedrich Abegg's travel diary, Frieder Otto Wolf: Art. 'Humanismus', *Historisch-kritisches Wörterbuch des Marxismus*, 6.1 (2004), 548–53 (p. 549).

33 Art. 'Humanität', in: Johann Heinrich Zedler, *Großes vollständiges Universallexikon*, 13 (1735), p. 1156; Stengel, *Genealogie*, p. 26.

direction: 'communicability and urbanity'[34] are linked to humanity [Humanität] as bourgeois etiquette, in the *Metaphysik der Sitten* [Metaphysics of Morals]: *compassibilitas.*[35] Later he distinguishes between *humanitas practica,* which refers to the ability and the willingness to communicate 'feelings' with one another, from *humanitas aesthetica,* or man's 'receptivity to the shared experience of pleasure or pain'.[36] However, for Kant, humanity is neither a historical-theoretical concept nor an educational ideal. This is similarly the case with other authors who, like Jakob Brucker[37] or Johann Gottfried Herder,[38] did not associate *humanitas* with the 15th and 16th centuries or with antiquity.

5.2. *Hegel?*

Hegel does not use the term 'humanism' either; instead, he rarely employs 'Humanität' in a manner similar to Kant or as the substantial aspect of humanity. He even rejects humanity as an educational project because it would contradict the self-realization of the universal master artificer. Likewise, the idea of humanity is only

34 Immanuel Kant: *Logik*, in: Immanuel Kant, *Gesammelte Schriften*, ed. by Königlich Preußische Akademie der Wissenschaften et al., 29 vols (Berlin: Reimer et al. 1902–2022), cit. as AA = Akademie-Ausgabe, AA 09, p. 46. For Kant cf. Stengel, *Genealogie*, 26–8.

35 Kant, AA 06, p. 335. Compassibilitas is *Hapax legomenon* in Kant, further proof cannot be provided at the moment. It is conceivable that Kant came up with the concept of humanity through his preoccupation with Herder's *Ideen zur Philosophie der Geschichte der Menschheit* [Ideas for the Philosophy of a History of Mankind]. The term appears eight times in Kant's review, notably also in the understanding of religion as the highest form of humanity. Kant, AA 08, pp. 45–52 (p. 49).

36 Kant, AA 06, p. 456.

37 Cf. Johann Jacob Brucker, *Kurtze Fragen aus der philosophischen Historie, von Christi Geburt biß auf unsere Zeiten*, 7 vols (Ulm: Bartholomäus 1731–7), vol. 6 (1735), 1, 4–5; vol. 5 (1734), 1327–8.

38 Cf. in particular Johann Gottfried Herder, 'Briefe zu Beförderung der Humanität', in: Johann Gottfried Herder, *Gesammelte Werke*, ed. by Hans Dietrich Irmscher (Frankfurt/Main: Deutscher Klassiker Verlag), vol. 7 (1991).

tangentially present in Hegel. A 'spirit of humanity' in Niethammer's sense would be rejected – from the perspective of world spirit.

Up to that point, receptions in the educational field prevail, now with the counter-concept of realism or naturalist education. Niethammer had still assigned a national and universal key function to humanism, which some educators expanded around 1830 by introducing 'humanism with a Christian and national orientation', thus incorporating the religious emphasis of Christianity – complementary, not contradictory.

5.3. *Humanism and 'Christianism'*

Christianismus, Humanismus und Rationalismus in ihrer Identität [Christianism, Humanism, and Rationalism in their Identity] is the title of a work from 1831 discussing Luther's Reformation ideas.[39] His name now appears for the first time in the humanism debate – linking back to the 16th century. Here, Luther is presented as a reconciler of the essence of humanity and the essence of Christianity. Notably, at the same time, there is a dispute about Luther's relationship with Erasmus, who later became the main representative of humanism with his *De libero arbitrio* (1524) against Luther's determinism, in front of *De servo arbitrio*, Luther's harsh writing against the assertion of human free will by Erasmus. The two texts only later became the prototypical expression of the clash between humanism and the Lutheran Reformation. But in Leopold von Ranke's monumental history of the Reformation, a conflict between Luther and Erasmus was simply not mentioned, just as it is absent in Niethammer's edition of Luther from 1817 and in most confessional editions of Luther that often do not even include until the 1840s. Those who wanted to adapt Luther to contemporary liberalism obviously wanted to conceal Luther's unfree man from Erasmus' *humanitas christiana*.[40]

39 Wilhelm Schröter, *Christianismus, Humanismus und Rationalismus in ihrer Identität: Ideen zur Beurtheilung der Reformation Luthers und des in ihr wahrhaft Symbolischen* (Leipzig: Vogel, 1831).

40 Stengel, *Genealogie*, pp. 34–6.

5.4. *Humanism as an Epoch Beside the Reformation: Hagen's Humanism*

As the first documented instance, in 1841, the radical democratic historian and member of the National Assembly in 1848, Karl Hagen,[41] regarded humanism as a movement since the 15th/16th century and as a distinct current alongside the Reformation. In his history of the Reformation (1841–1844), published at the same time as Ranke's opus magnum and Wilhelm Zimmermann's mammoth history of Peasants' War,[42] however, humanism is presented as a Christian alternative and is identical to modern science; it is merely a component of the opposition against Rome and scholasticism. He identifies three strands of opposition: popular, religious, and humanistic, that gradually converged, mainly on issues relating to the ecclesiastical situation, Rome and scholasticism.[43] Luther was in the opposition grouped with Erasmus, Reuchlin, Pico della Mirandola and others and they effectively united the popular, humanist and religious adversaries.[44] But in the second volume of his History of Reformation, Hagen made a strong distinction between Protestantism and the 'original tendency of the Reformation'.[45] For Hagen, the Protestantism is an 'deviant variant [Abart]' of the essence of the Reformation; as orthodoxy, it stands opposed to humanism, with both having gone their separate

41 Cf. Günter Mühlpfordt, 'Karl Hagen: Ein progressiver Historiker im Vormärz über die radikale Reformation', *Jahrbuch für Geschichte* 21 (1980), pp. 63–101; Eike Wolgast: 'Karl Hagen in der Revolution von 1848/49: Ein Heidelberger Historiker als radikaler Demokrat und politischer Erzieher', *Zeitschrift für die Geschichte des Oberrheins*, N.F. 94, 133 (1985), pp. 279–99.

42 Wilhelm Zimmermann: *Allgemeine Geschichte des grossen Bauernkrieges: nach handschriftlichen und gedruckten Quellen*, 3 vols (Stuttgart: Köhler, 1841–3, many edn, recently 11th edn, Berlin (GDR) 1989).

43 Karl Hagen, *Deutschlands literarische und religiöse Verhältnisse im Reformationszeitalter: mit besonderer Rücksicht auf Wilibald* [sic!] *Pirckheimer*, 3 vols (Erlangen: Palm, 1841–4), vol. 1 (1841), p. 32; Stengel, *Genealogie*, pp. 36–43.

44 Hagen, vol. 1 (1841), pp. 232–3, 256, 464.

45 Hagen, vol. 2 (1843), p. x.

ways ever since.[46] For the Hegelian Hagen, humanism became a spirit that could temporarily federate with other spirits but ultimately conveyed 'ideals of freedom and humanity'[47] that had been lost in the political situation of the Vormärz, the years before the 1848 revolution. To create a common front in the republican sense, Hagen may have imported the ideological front lines from the 1840s into his analysis. Until the historiographies of the 21st century, the humanism group oscillates between 'humanists', Hermetics, Spiritualists, and especially Anabaptists, from Erasmus, Agrippa, Miguel Servetus to the Swiss Anabaptists. The three volumes present the pinnacle and the culmination of the Reformation movement, with 'Sebastian Franck, the forerunner of modern German philosophy',[48] who was the first translator of Corpus Hermeticum into German.[49] The democratic opponent Hagen apparently regarded him as his spiritual ancestor.

By the way, Hagen was active in reform education through his connections to the reform pedagogues Friedrich Fröbel and Johann Heinrich Pestalozzi,[50] eventually advocating for free schools and propagating 'humanism' as a means to educate and engage in political critique.

Overall, humanism for Hagen is Christian, educational and a being that has become history and continually acts through history, primarily in relation to the Reformation – first symbiotically and then in conflict. Thus, it has entered into the interpretative horizons of historians as a historical force and as a Christian-liberal alternative against Lutheran state Protestantism.

---

46 Hagen, vol. 3 (1844), p. vii.

47 Hagen, vol. 3 (1844), p. 458.

48 Hagen, vol. 3 (1844), pp. 314–96.

49 Kristine Hannak, 'Pymander als inneres Wort. Sebastian Francks Übersetzung des *Corpus Hermeticum* in der Tradition mittelalterlicher Logosmystik', in: *Konzepte des Hermetismus in der Literatur der Frühen Neuzeit*, ed. by Peter-André Alt and Volkhard Wels (Göttingen: V&R Unipress, 2010), pp. 297–321.

50 Cf. Karl Hagen, 'Über Nationale Erziehung: Mit besonderer Rücksicht auf das System Friedrich Fröbels [1845]', in: Friedrich Fröbel and Karl Hagen, *Ein Briefwechsel aus den Jahren 1844–1848*, ed. by Erika Hoffmann (Weimar: Werden und Wirken, 1948), pp. 97–136 (p. 99–100); Stengel, *Genealogie*, 42–3.

5.5. *Humanism and Communism: Just an Interlude?*

It seems that Marx and Engels briefly arrived at humanism through Karl Hagen. 1844, in *Die Heilige Familie* [The Holy Family] and in the Ökonomisch-philosophische Manuskripte [Economic and Philosophic Manuscripts], Marx explicitly positioned humanism against contemporary 'spiritualism' as its most dangerous enemy and in a materialistic manner: 'Real humanism has no more dangerous enemy in Germany than spiritualism or speculative idealism, which substitutes "self-consciousness" or the "spirit" for the real individual and teaches with the evangelist: "It is the spirit that makes alive, the flesh is of no use". It is understood that this fleshless spirit has spirit only in his imagination'.[51]

In his *Economic and Philosophic Manuscripts* Marx linked humanism to atheism and communism: atheism is the 'coming of age of theoretical humanism', while communism is the 'coming of age of practical humanism'.[52] Here, humanism is at the center of a human-centered philosophy and simultaneously part of a program that is not only critical of the church but also atheistic. But the claim of some authors, including Michael Schmidt-Salomon, that Marx had called for a 'categorical imperative of humanism', has been systematically refuted as an erroneous conflation of Kant's moral philosophy with Marx's short-lived use of the concept of humanism.[53]

However, only four years later, in 1848, in the *Manifest der Kommunistischen Partei* [Communist Manifesto], humanism was sharply rejected. Humanists were now explicitly counted among the movements, along with philanthropists and spiritists ('a specter is haunting…'), that sought to secure the existence of bourgeois society.[54]

51 John 6,63; Karl Marx and Friedrich Engels, 'Die heilige Familie' (September 1844), in: Karl Marx and Friedrich Engels, *Werke*, 44 vols (Berlin: Dietz, 1956–2018), cit. as MEW = Marx–Engels–Werke, MEW 02, pp. 3–223, (p. 7, Vorrede/Preface).

52 Karl Marx and Friedrich Engels, *Gesamtausgabe* (Berlin: Akademie der Wissenschaften, 1975–), cit. as MEGA = Marx–Engels–Gesamtausgabe, MEGA I.2, p. 301 (Erste Wiedergabe).

53 Baab, pp. 159, 191.

54 MEW 04, pp. 461–93 (p. 488).

In *Die großen Männer des Exils* [Heroes of the Exile, 1852], Marx and Engels settled accounts with former left Hegelian partisans, who had tried to save humanism: 'the phrase that all confusionists in Germany, from Reuchlin to Herder, have used to cloak their embarrassment'.[55]

As a result, humanism disappeared from the Marxist and socialist movement until the Popular Front movement of the 1930s,[56] before the GDR adopted it as part of its dazzling vocabulary of the 'church struggle' – i.e. opposed to Christian humanism – or as a compromise term to make a hegemonic claim to a common denominator between Marxists and Christians.[57] What lay behind Marx's and Engels' rejection of humanism as a quasi-religious worldview?

5.6. *Splits among the Left Hegelianisms: Humanism as Lodge, Religion, and Invisible Church*

Marx's opponents were two former young Hegelian allies, Moses Hess[58] and, in particular, Arnold Ruge. Like Karl Hagen, Ruge was a member of the Frankfurt National Assembly in 1848 and represented the radical-democratic Reform Party. Already in 1826, he was

---

55 MEW 08, pp. 233–335 (p. 278); Baab, p. 45.

56 Cf. Walter Rüegg, Art. 'Humanismus', II. Philosophisch', RGG, 3rd edn, 3 (1959), pp. 479–82 (p. 480).

57 Hartmut von Hentig reports on a humanism conference in Wittenberg in 1961, at which humanism was not defined, even once, and the humanism of the young Marx was not mentioned by the GDR scholars present, although it was SED's official term paralleling socialism. Of all people, it was a West German who pointed to the currency of humanism in Marx, cf. Hartmut von Hentig, 'Humanismus und die DDR: Von der Ohnmacht des Namens', *Frankfurter Hefte*, 16.2 (1961), 81–92 (p. 85). This is remarkable in view of the fact that the concept of humanism was experiencing a boom in the state and party leadership of the GDR since October 1960. On 4 October 1960, Walter Ulbricht, leader of the SED, claimed in a programmatic declaration: 'Christianity and the humanistic goals of socialism are not opposites'. Cf. Marxisten und Christen wirken gemeinsam für Frieden und Humanismus, ed. by Staatsrat der Deutschen Demokratischen Republik (East Berlin: Staatsverlag, 1964), pp. 85–6. On that connections cf. in detail now Blume, 2021.

58 Cf. Stengel, *Genealogie*, p. 47.

imprisoned and later went into exile; he was a philosopher habilitated on Plato and also studied under Friedrich Schleiermacher, whose imprint is unmistakable in his writings.[59] Ruge's programmatic text *Die Religion unserer Zeit* [The Religion of Our Time] was primarily shaped by Feuerbach's critique of church and religion but strictly adhered to the concept of religion, asserting that religion is the 'heartbeat of the moral world'.[60] In the Reformation, in the socialism of the revolution, and in the 'seriousness of enlightenment', Ruge saw 'real additional developments of the Christian principle of humanity'.[61] In this 'humane religion', the idea of God was fulfilled and transcended in the man himself. This could be described as a secularized Christology of incarnation. However, Ruge did not connect anywhere to the Renaissance; he saw no difference between the Reformation and 'humanism'.

In his programmatic pamphlet *Die Loge des Humanismus* [The Lodge of Humanism], published in 1852, and then in his *Reden über Religion ihr Entstehen und Vergehen an die Gebildeten unter ihren Verehrern* [Speeches on Religion, Its Emergence, and Decay Dedicated to the More Educated Among His Followers] of 1869, the 'lodge of humanism' was the 'invisible church of mankind'.[62] For the more human the god of a religion is, the truer he is.[63] But the entire Christian heaven must fall down at the moment the true nature of Christianity is recognized. Then a 'religion, the humane one' would emerge, where there would no longer be otherworldly

59 On Ruge cf. an earlier work of Margarete Pohlmann, *Der Humanismus im 19. Jahrhundert – Eine neue Religion? Arnold Ruges Auseinandersetzung mit dem Christentum* (Frankfurt/Main, Bern and Las Vegas: Lang, 1979), as well as Stefan Walter, *Demokratisches Denken zwischen Hegel und Marx: Die politische Philosophie Arnold Ruges. Eine Studie zur Geschichte der Demokratie in Deutschland* (Düsseldorf: Droste, 1995); Baab, pp. 38–42; Stengel, *Genealogie*, pp. 49–57.

60 Arnold Ruge, *Die Religion unserer Zeit* (Leipzig: Vereins Buchdruckerei, 1849), p. 10.

61 Ruge, *Religion*, pp. 13–4.

62 Arnold Ruge, *Die Loge des Humanismus* (sine loco: Verlag des Herausgebers, 1852), p. 9.

63 Cf. Ruge, *Loge*, p. 19.

gods but only living humans.[64] The transcendent God becomes this world; the new religion is a result of interpreting old Christianity.

Ruge's humanism thus became for the first time an independent socio-political program, as a continuation of the Christian idea of humanity directed against the established church but not against Christianity itself. Ruge understood humanism as a socialist and simultaneously democratic alternative to Marxist program, as a fulfillment of Christianity and in clear opposition to state Protestantism, yet as a religion-like program that was not simply pedagogical and also transformed immortality into an immanent process with complete incarnation of God in humanity at its end. For Marx and Engels, Ruge's interpretation was evidently the reason to reject humanism as an idealistic-revisionist project, as I mentioned above: 'Then Ruge saved himself behind humanism, that phrase with which all confusionists in Germany, from Reuchlin to Herder, have used to cloak their embarrassment', they wrote in 1852.[65]

Some claim that Ludwig Feuerbach is one of the most important authors for the conceptualization of humanism. But a closer examination reveals the opposite: humanism occupies only a marginal place in Feuerbach's work, sometimes as a parallel to rationalism in the sense of a liberal interpretation of the Bible or as the 'deification of the real, of what materially exists', as 'theological atheism', 'pantheism', or 'materialism'.[66] Feuerbach's reinterpretation of theology in anthropological terms might have been classified as humanism by those around him, and later by Arnold Ruge, but certainly not by Feuerbach himself. Thus, Feuerbach's contribution to the debate on humanism would be merely the history of the impact of his recipients.

---

64 Cf. Ruge, *Loge*, p. 25.

65 Marx and Engels, 'Die großen Männer des Exils' (1852), MEW 08, pp. 233–335 (p. 278); Baab, p. 45.

66 Cf. Ludwig Feuerbach, 'Grundsätze der Philosophie der Zukunft', in: Ludwig Feuerbach, *Sämtliche Werke*, 2nd edn (Stuttgart–Bad Cannstatt: Frommann–Holzboog, 1959–60), vol. 2, pp. 264–5. In addition to that, 'humanism' appears only in the posthumous aphorisms, cf. ibid., vol. 10, p. 300; Stengel, *Genealogie*, pp. 57–9.

5.7. *Humanism as the Spirit of Antiquity in Opposition to the Reformation and Christianity*

The decisive anti-church and simultaneously historiographical content of humanism reached its peak through the historian Georg Voigt.[67] His *Wiederbelebung des classischen Alterthums oder das erste Jahrhundert des Humanismus* [Revival of Classical Antiquity or the First Century of Humanism], is regarded as a standard work, much like Jacob Burckhardt's *Kultur der Renaissance in Italien* [The Civilization of the Renaissance in Italy], first published in 1860. Burckhardt mentioned humanism somewhat incidentally. His treatment of humanism was rather casual, he did not define it and simply used it to analyze Pico della Mirandola and humanism's as an educational movement, critical to the church, which had been defeated by the 16th century.[68]

In the first edition of his *Wiederbelebung des classischen Alterthums oder das erste Jahrhundert des Humanismus* from 1859 Georg Voigt did not position humanism in opposition to Christianity; it was merely a 'revival of classical antiquity'.[69] Without any comment or explanation, it was only in the second edition in 1880 that Voigt added that the views of humanism were 'opposed to the views of Christianity and the church'.[70] He stated that humanism had been 'undoubtedly a natural born enemy of the church'[71] and had waged a 'ruthless war [...] against astrologers,

67 On Voigt cf. Paul F. Grendler, 'Georg Voigt: "Historian of Humanism"', in: *Humanism and Creativity in the Renaissance*: Essays in Honor of Ronald G. Witt, ed. by Christopher S. Celenza und Kenneth Gouwens (Leiden: Brill, 2006), pp. 295–326.

68 Cf. Jacob Burckhardt, *Die Kultur der Renaissance in Italien: Ein Versuch* (Frankfurt/Main: S. Fischer, 2009 = 2nd edn, 1869), pp. 191–215; Stengel, *Genealogie*, pp. 63–4.

69 Georg Voigt, *Die Wiederbelebung des classischen Alterthums oder das erste Jahrhundert des Humanismus*, 3rd edn, 2 vols (Berlin: Reimer 1st edn 1859, 2nd edn 1880–1, 3rd edn 1893), 1859, p. 4 = 1880, vol. 1, p. 3.

70 Voigt, 1880, vol. 1, p. 4.

71 Voigt, 1881, vol. 2, p. 213. The section 'The Humanism and the Church' is missing at the end of the 5th chapter in the 1st edn (1859).

alchemists, dream interpretation and superstition in all its forms'.[72] This assertion cannot be upheld, especially for the Hermeticists and Kabbalists Pico, Ficino, and Reuchlin, who were still counted among the humanists by their contemporary Jacob Burckhardt. However, Voigt's chain of reasoning is significant: the humanists were the first 'apostles of enlightenment',[73] they were thus non-Christian carriers of science and education for progress. In 1859, the humanists had aimed to 'recapitulate antiquity *and* the pinnacle of Christian-romantic life'.[74] There was no hostility between them. There was no contradiction between the two in this instance; they simply needed to be summarized and reiterated.

In 1880, this passage is absent. Now, Voigt's humanism was not only anti-church, but also anti-Christian. The anti-Christian line runs from antiquity through the Renaissance to the Enlightenment – and into contemporary times, insofar as it concerns the unfolding of a transhistorical spirit.

At the moment we cannot say exactly what caused this radical change, but it was probably due to the 'Kulturkampf' [cultural struggle] of the Prussian state under Reichskanzler Bismarck against Roman Catholicism in Germany under the heading of ultramontanism in the 1870s. Only some years later, in 1885, Voigt's definition of humanism in contrast to Christianity and with origin in the classical antiquity is established in the most influential encyclopedias: humanism is now defined as a literary 'movement' related to antiquity which originated in classical antiquity and is contrasted with the 'Christian-ecclesiastical principle of life of the Middle Ages', whereby the 'principle of life' is also that of 'humanity'.[75] The fundamental conflict between Christianity/the church and humanism, which Voigt emphasizes, is, however, simply attributed to the Middle Ages, and this limitation is further applied to the German humanists with the justification that humanism was an intermediate stage in the development between the ancient life

72 Voigt, 1893, vol. 2, p. 486.

73 Voigt, 1893, vol. 2, 486 = 1881, vol. 2, p. 492.

74 Voigt, 1859, p. 3 [author's emphasis].

75 Ernst Götzinger, Art. 'Humanismus', *Reallexicon der Deutschen Altertümer*, 2nd edn (Leipzig: Urban, 1885), pp. 435–40 (p. 435).

principle of '*humanitas*' and the 'humanitarian ideals of the 18th century'[76] – without explicitly using the term 'Enlightenment'. At the core lies the 'rebirth of the ancient world view',[77] which is reminiscent of the ability ascribed to spiritual beings to reincarnate – in the contemporary reincarnation discourse and in the worldview of philosophical contemporaries.[78]

An excellent example of how context produces and generates science.

6. *Humanism – Not a Viable Historiographical and Heuristic Term!*

It cannot even be said in the slightest that the discussion has been concluded. Without repeating the arguments in detail: neither the opposition to the church and Christianity can be upheld for all humanists, nor are the references to antiquity, philosophy, language, and the immortality of the soul clear and uniform. Humanism shimmers between fronts as a pedagogical, philological, (alternative) religious, socio-political term, and as a designation of a superhistorical being or spirit that cannot be reduced to a common denominator without ignoring historical contexts. My examination stops with Voigt in the 1880s because here the coordinates for the concept of humanism exist between education, religion, history, anthropology, and worldview, in which the term has since been used depending on context and positioning. This finding already raises doubts about whether it is even possible to use the concept of humanism as an 'indispensable heuristic tool'[79] for historical-critical work and in this sense for contemporary discourse. And it is

---

76 Götzinger, p. 435.
77 Götzinger, p. 439.
78 Cf. on the debates about Alan Kardec's Buch der Geister (1857) and, for example, the early socialist Charles Fourier, briefly Diethard Sawicki, *Leben mit den Toten: Geisterglauben und die Entstehung des Spiritismus in Deutschland 1770–1900*, 2nd edn (Paderborn et al.: Schöningh, 2016), pp. 283, 287–96.
79 See above, note 30; Walther, p. 668.

a mystery how one could use humanism as a valid historiographical category with universal and transhistorical scope.

The simplest objection would be that the *humanista* of the 14th to 16th centuries were indeed 'only' language teachers and students of the *humaniora* without a common anthropological agenda. It is more challenging to establish a consistent position regarding the church, Christianity, or the Reformation. In some authors, one can recognize a critical stance towards church institutions but not necessarily a critical stance towards Christianity as such.

In contrast, there is a theological anthropology in the theological-philosophical debate that stands against Luther's radically anti-Pelagian Augustinian interpretation as an Augustinian Hermit and favors Erasmus' Augustinian interpretation as an Augustinian Canon who grants free will its own significance even in matters of salvation – similar to the followers of Philipp Melanchthon, later Pietists, or Roman Catholics. However, Luther's understanding of Augustine has been contested from the very beginning within Lutheranism and is by no means consensual.[80]

7. *No 'Humanism' in the Early Modern Times*

From the outlined history of the invention of humanism, it has become clear that humanism has always been used as a term of confrontation or alternative. But humanism also functions as a concept of concealment or veiling. To this end, we can consider a case from the central question of anthropology: Typically, Giovanni Pico della Mirandola and Marsilio Ficino are cited as reference authors for early modern humanism.[81] When one inquires about the new, anthropocentric image of humanity presented by these authors, one quickly encounters the widely discussed speech on human dignity by Giovanni Pico della Mirandola, *Oratio de*

80 See above section 5.3; Stengel, *Genealogie*, pp. 76–9.

81 With the exception of Paul Oskar Kristeller, cf. Paul Oskar Kristeller, *Humanismus und Renaissance*, ed. by Eckhard Keßler, 2 vols (München: Fink, 1980), vol. 1, pp. 58–61, see also vol. 2, p. 249 as well as pp. 11–29, vol. 2, 244–64.

hominis dignitate. If the 'anthropocentrism of the humanistic conception of man' is viewed as a negative precondition for Luther's Reformation,[82] then the *Oratio* is where this conception of man is most clearly and influentially represented. In the *Oratio*, it is initially impossible to find an anti- or un-Christian accent. It represents a novel Christian theology influenced by Hermeticism and Christian Kabbalah, whose founders were Ficino for Hermeticism and Pico for Christian Kabbalah. Their texts have been extensively received by later so-called humanists such as John Colet, Thomas More, Erasmus, Reuchlin, Andreas Osiander, up to Sebastian Franck,[83] but also by prominent opponents of Luther like his most famous opponent Johannes Eck,[84] and Hieronymus Emser, who published Pico's *Opera omnia* in 1504, the same year Luther was as Emser's student in Erfurt familiar with Johannes Reuchlin's *Sergius*[85] and, I argue, began to develop his deep aversion to hermetic and cabalistic soteriology and anthropology.[86] Because Pico was the subject of official church proceedings, his texts were often referred to, but his name was not mentioned, especially in the case of Erasmus.

But what are the roots of Pico's image of humanity? It lies in the cosmic anthropology and angelology of the Muslim philosopher

82 Cf. Thomas Kaufmann, *Geschichte der Reformation* (Frankfurt/Main, Leipzig: Verlag der Weltreligionen im Insel-Verlag, 2009), p. 123.

83 See above note 49.

84 See Giovanni Tortoriello, 'The Prisca Theologia in the Early Reformation Debates', *Renaissance and Reformation / Renaissance et Réforme*, 47.2 (2024), 41–71; Id., 'The Transformations of "Renaissance Aristotelianisms": The Case of Johannes Eck's Commentary to the *Corpus Aristotelicum*', in: *Reformation & Renaissance Review* [henceforth RRR], 25 (2023), 63–81.

85 Cf. Friedemann Stengel, 'Reformation, Renaissance and Hermeticism: Contexts and Interfaces of the Early Reformation Movement', *RRR*, 2 (2018), 103–33 (pp. 107, 113–15); Id., 'Reformation, Renaissance und Hermetismus: Kontexte und Schnittstellen der frühen reformatorischen Bewegung', *Archiv für Reformationsgeschichte*, 104 (2013), 35–81 (pp. 37, 51).

86 Now as groundbreaking and comprehensive exposition of this line of conflict: Giovanni Tortoriello, *Scala Christus est: Reassessing the Historical Context of Martin Luther's Theology of the Cross* (Tübingen: Mohr Siebeck, 2023).

Algazel, derived from Hermes Trismegistus's angelic world combined with an anthropocentrism compiled from Kabbalistic sources. This anthropocentrism sees Adam Kadmon as the godlike primordial man and the center of all existence, endowed with free will and agency. He is a prototype and paradigm of humanity capable of deciding whether to ascend beyond the angelic realm into the divine sphere or to become beast-like.[87]

For Pico – whom Erasmus quotes almost verbatim in his *Enchiridion militis christiani* only nine years after Pico's death[88] – the human being is neither earthly nor heavenly but rather a divine being connected to the 'secret theology' of the Hebrews and Pythagoreans. Pico's humans stand in the 'center of the world' and possess more potential than the highest spirits.[89] For Ficino, in his *Theologia platonica*, they are 'Dei vicarii in terra'[90] – the Vicarius Christi remains unnamed. The human is modeler and maker (plastes et fictor) of himself.[91]

This anthropocentrism is profoundly theological. In casting humans as having been created first in the image of God, the linking and not the separating factors come to the fore. But, nonetheless, it has significant consequences for the role of Jesus Christ and the church in the justification process: neither necessary sacraments for salvation nor priests nor church nor pope appear in this anthropology.

---

87 Giovanni Pico della Mirandola, *De hominis dignitate: Über die Würde des Menschen* (Hamburg: F. Meiner, 1990), pp. 6–7. For more on that: Walter Andreas Euler, *'Pia philosophia' et 'docta religio': Theologie und Religion bei Marsilio Ficino und Giovanni Pico della Mirandola* (München: Fink, 1998), pp. 99–122, on Ficino ibid., pp. 259–68.

88 Erasmus von Rotterdam, 'Enchiridion militis christiani: Handbüchlein eines christlichen Streiters', in D. Erasmus von Rotterdam, *Ausgewählte Schriften*, ed. by Werner Welzig, vol. 1 (Darmstadt: Wissenschaftliche Buchgesellschaft, 1995), pp. 56–375 (pp. 109, 143); Stengel, Hermetismus, pp. 55–6.

89 Pico, pp. 5–9.

90 Ficino, *Theologia Platonica* XVI 6–7, as cited in Jörg Lauster, *Die Erlösungslehre Marsilio Ficinos: Theologiegeschichtliche Aspekte des Renaissanceplatonismus* (Berlin, New York: De Gruyter, 1998), pp. 58, 62.

91 Pico, p. 6.

Moreover, the cross and the attribution of Christ's merits for the salvation of believers are replaced by a soteriology directed towards resurrection – even if there is no salvation without grace according to Pico, Ficino and his followers.[92]

The human conception presented by Pico, Ficino, and subsequently Erasmus and Reuchlin stands in opposition to a *specific* understanding of Christianity that is Augustinian in nature and vehemently defended by Luther with a clear anti-Pelagian soteriology and what could be described as a – without justification through Christ – very pessimistic anthropology. Recently, it has been convincingly shown by Giovanni Tortoriello that young Luther's protest was directed precisely against this anthropology of Ficino and Kabbalah as represented by Reuchlin and Pico, where the cross as an act of redemption was transformed into a theology of overcoming the dead and immortality. As I mentioned, Luther's arch-enemy, his former professor at Erfurt, Hieronymus Emser, was one of the editors of Pico's *Opera omnia*. Just one year before the 95 Theses were published, Johannes Eck released a commentary on Ficino's hermetic anthropology and angelology in his *Mystical Theology*, and published the hermetic text *Chrysopassus* in 1514.[93] For the radical cross-centered Christocentric Luther, this was blasphemy; however, it was widespread within the church. Pico's anthropology modified soteriology but remains Christocentric nonetheless.[94]

If this human conception is categorized under the label of humanism – as is often done – it risks creating conceptual confusion because, *first*, it understands itself as Christian. *Second*, this

---

92 Cf. Lauster, pp. 82–4, 110–2, 121–2, 124–56 and more. For Ficino and Pico, the generalizing statements about the allegedly completely un- or anti-soteriological theories of perfectibility and the anthropocentricity of 'humanism' are to be questioned. Cf., for example, Harm Klueting, *Das Konfessionelle Zeitalter: Europa zwischen Mittelalter und Moderne. Kirchengeschichte und Allgemeine Geschichte* (Darmstadt: Primus, 2007), p. 101.

93 Tortoriello, Scala, pp. 184–7.

94 Stengel, Hermeticism, pp. 108, 124, 129–30; Stengel, Hermetismus, pp. 36, 42–4, 70, 79–81.

humanism relates to *studia humanitatis* insofar as it does not merely translate and integrate receive Plato, Plotinus, and Aristotelians but also incorporates hermetic and Kabbalistic writings along with demonological, alchemical, and astrological texts within a framework of *philosophia christiana*.[95]

Undeniably *thirdly* Pico's Kabbalistic-hermetic anthropology – and that of Ficino and their recipients – is naturally critical of the church because particularly the sacramental mediating role of priests and sacraments along with that of *vicarius Christi* no longer appears or is marginalized. These roles had been dogmatized under Innocent III in 1215, some years after his *De miseria humanae conditionis* was directed against Pico's significant predecessor Giannozzo Manetti with his *De dignitate et excellentia hominis*.[96]

95 Ficino's collection of twenty demonological-Hermetic essays was published for the second time in Venice in 1516 with a dedication to Leo X.: Cf. the copy in the Halle University Library, which has been worked through intensively and has numerous underlinings and an ownership entry from a Magdeburg high school: *Iamblichus de mysteriis Aegyptiorum. Chaldeorum. Assyriorum. Proclus Platonicum Alcibiadem de anima, atq. daemone. Proclus de sacrificio & magia. Porphyrius de divinis atq. daemonibus. Synesis Platonicus de somnis. Psellus de daemonibus. Expositio Prisciani & Marsilii in Theophrastum de sensu, phantasia & intellectu. Alcinoi Platonici philosophi liber de doctrina Platonis. Speusippi Platonis discipuli liber de platonis definitionibus. Pythagorae philosophi aurea verba. Symbola Pithagorae philosophi. Xenocratis philosophi platonici liber de morte. Mercurii Trismegisti Pimander. Eiusdem Asclepius. Marsilii Ficini de triplici vita Lib. II. [...]*, 1516.

96 Lotario de Segni (Pope Innocent III), *Vom Elend des menschlichen Daseins*, trans. by Carl-Friedrich Geyer (Hildesheim, Zürich, New York: Olms, 1990); Walter Wili, 'Innozenz III. und sein Werk "Über das Elend des menschlichen Daseins"', in: *Humanismus, Mystik und Kunst in der Welt des Mittelalters*, ed. by Josef Koch (Leiden, Köln: Brill, 1953), pp. 125–361; Giannozzo Manetti, Über die Würde und Erhabenheit des Menschen: De dignitate et excellentia hominis (Hamburg: F. Meiner, 1990); on Manetti: Charles Trinkaus, *In Our Image and Likeness: Humanity and Divinity in Italian Humanist Thought* (Notre Dame, IN: University of Notre Dame, 2012), pp. 230–70, 578–601.

8. *New Perspectives on Engaging with Humanism*

How should Pico's and Erasmus' anthropology form even a heuristic, historiographical, let alone normative category when compared to Niethammer's pedagogical approach, Marxism – which rejects metaphysical explanations of humanity – or Ruge's secular 'lodge church' of humanism, not to mention socialist humanism in the GDR? To adequately address both the clearly theological content and the specific nature of the so-called appropriation of antiquity, I find it appropriate to speak of 'Christian Hermeticism' or 'Christian Kabbalah' instead of humanism for the early modern period.[97] A precise historical perspective prevents the compulsion to adapt early modern paradigms from the 19th and 20th centuries.

I disagree with certain researchers of esotericism who advocate using the term 'esotericism' instead, in part because it originated only in the 19th century, and notwithstanding its previous sporadic use, it would be imposed on authors of the early modern period. That would undoubtedly lead to the same issues as 'humanism'.[98] This would, for example, amplify a deviant religious and scientific current in Ficino and Pico which runs even counter to established Christianity, even though these authors clearly identified and

97 Cf. Stengel, *Genealogie*, pp. 79–83.

98 In my view, the fact that the term 'esotericism' has been used in German-speaking countries since the 18th century in connotation with Masonic and Pythagorean currents does not make it any easier to look at it beyond these concrete contexts as a suprahistorical movement since the early modern period. Cf. on that Monika Neugebauer-Wölk, 'Historische Esoterikforschung, oder: Der lange Weg der Esoterik zur Moderne', in: *Aufklärung und Esoterik: Wege in die Moderne*, ed. by Monika Neugebauer-Wölk, Renko Geffarth and Markus Meumann (Berlin, Boston: De Gruyter), pp. 37–72; Friedemann Stengel, 'Lebensgeister – Nervensaft: Cartesianer, Mediziner, Spiritisten', in: ibid., pp. 340–377 (pp. 345–348); Stengel, *Enlightenment*, pp. 832–7; Michael Bergunder, 'Was ist Esoterik? Religionswissenschaftliche Überlegungen zum Gegenstand der Esoterikforschung', in: *Aufklärung und Esoterik: Rezeption – Integration – Konfrontation*, ed. by Monika Neugebauer-Wölk and Andre Rudolph (Tübingen: Niemeyer, 2008), pp. 477–507.

regarded themselves as Christian.[99] They represented a theology that differed from only the anti-Pelagian Augustinian Christianity, but it was essentially Christian, with the above-mentioned shifts in soteriology and anthropology. One could even describe Erasmus as being strongly Christocentric.

From a historical perspective, it should not be determined *post res* which form of Christianity is 'true', 'orthodox' or 'real' to be able to distinguish it from esotericism, heterodoxia, or even humanisms. What constitutes 'orthodox', 'true', or lawful must be debated and is not a heuristic presupposition of a historical perspective. It would be historically appropriate to stay at the respective contextual level – knowing that concepts and receptions as well as theosophical designs and alchemical practices pertaining to Kabbalah and Hermeticism often overlap among early modern authors. Yet it becomes implausible to subsume these currents under concepts that emerged in the 19th century. A clear historical perspective, on the other hand, will prevent the early modern debates from being modified to fit into specific 19th- or 20th-century paradigms.

9. *Conclusion: Enlightenment as Criticism – also of 'Humanism'!*[100]

As a historiographical and heuristic category, the concept of humanism is problematic because it cannot be freed from its heterogeneous and polygenetic contexts. It is also problematic in another respect, namely as a normative that is associated with certain human conceptions and is intended to provide guidance for action and reflection. Michel Foucault, in particular, has invigorated the debates surrounding the validity and normativity of a hegemonic concept of humanism. For Foucault, Enlightenment and humanism

---

99 Cf. in the case of Ficino, his work *De Christiana Religione*, which was always included in later editions of his works (e.g., in Marsilio Ficino: *Opera [...] una cum gnomologia*, 2 vols (Paris: Bechet 1641), vol. 1, pp. 1–73. Significantly, *De Christiana Religione* is followed in this complete edition by one of Ficino's main works entitled *Theologia Platonica de immortalitate animorum*, ibid.: pp. 74–414.

100 Cf. Stengel, *Genealogie*, 83–93.

are not closed systems that one merely needs to reconstruct in order to apply them, as if one only needed to 'correctly' understand a historical process in order to 'learn' from it for the future. Then, Enlightenment and Humanism would itself be attributed a hegemonic, heteronomy-promoting power, whereas, in accordance with Kantian critique, autonomy must also be achieved and preserved in relation to epistemological presuppositions. For Foucault, the Enlightenment is subject to historical critique as a concrete historical event, and humanism is critiqued because it is a theme embedded in specific discourses.[101] This stands in contrast to the principle of critique and the autonomous 'permanent creation of ourselves',[102] not as self-discovery in the sense of a metaphysical project returning to an alleged origin, but rather in terms of 'ascetic self-invention'[103] and an 'indeterminate work of freedom'.[104] It is not a specific image of humanity but rather the principle of critique that Foucault sees at the heart of the historical consciousness of the Enlightenment,[105] which reveals the historicity of images of humanity.

In the claim to use one's own reason to question the limits of one's existence, prevailing norms, and one's own possibilities for knowledge lies an ethos. Epistemology is linked with contemporaneity within an ethical stance aimed at changing the present.[106] Neither Enlightenment nor humanism should be regarded as extrinsic conditions of discourse; otherwise, they would hinder its analysis and perpetuate and universalize or eternalize historically constructed oppositions such as humanism versus non-humanism or anti-humanism. Criticism is the 'handbook' of reason that came

---

101 Michel Foucault, 'Was ist Aufklärung?', in: Michel Foucault, *Ethos der Moderne: Foucaults Kritik der Aufklärung*, ed. by Eva Erdmann, Rainer Forst and Axel Honneth (Frankfurt/Main, New York: Campus 1990), pp. 35–54 (pp. 46–7).

102 Foucault, Aufklärung, p. 47.

103 Foucault, Aufklärung, p. 45.

104 Foucault, Aufklärung, p. 49.

105 Foucault, Aufklärung, p. 47. Cf. also Michel Foucault, Was *ist Kritik?* (Berlin: Merve 1992); Judith Butler, 'Was ist Kritik? Ein Essay über Foucaults Tugend', *Deutsche Zeitschrift für Philosophie*, 50.3 (2002), 249–65.

106 Foucault, Aufklärung, pp. 45, 53.

of age in the Enlightenment, and the Enlightenment is the 'age of criticism'.[107] If humanism resists classification as a legitimate historiographical category for consolidating contemporary positions – that is, if it constitutes not a determinate phenomenon but rather a multi-stratified discursive arena – then the following theoretical propositions merit consideration:

1. Drawing upon Ernesto Laclau's political theory, I propose that humanism functions as an empty signifier – a discursive site wherein political negotiations unfold both within academic debates and populist mobilizations. Empty signifiers constitute contested terrains where the semantic filling of the signifier becomes simultaneously claimed and disputed. The signifier 'humanism' undergoes perpetual processes of semantic saturation while its particular content remains subject to ongoing contestation.

The impossibility of achieving universally accepted semantic closure reflects not a theoretical inadequacy but rather the inherent political nature of the signifier itself. Any ostensibly definitive filling could only be secured through what might be termed a 'violent closure' – a suppression of alternative semantic possibilities that forecloses the very polyvalence constitutive of democratic discourse. No transcendent arbitrator or referee exists who might adjudicate between competing definitions in the face of this fundamental semantic multiplicity.

2. In this theoretical framework, humanism constitutes a discourse whose epistemic coordinates and foundational assumptions become the object of critical interrogation centered upon a fundamental question: What constitutes the human and humanity, and how ought these categories be configured? This inquiry necessitates sustained analytical attention to the complex articulations between knowledge, power, and ethics that traverse debates concerning humanistic discourse. The powerful interconnections binding knowledge to power, and both to ethical frameworks, must remain transparent within our critical engagement with humanistic discourse. These connections do not merely influence the debate; they constitute its very possibility, shaping the parameters within which questions of

107 Foucault, Aufklärung, p. 41.

human essence and human flourishing can be meaningfully posed and contested.

3. Positioning oneself within this discourse necessarily implicates questions of reflexive self-constitution and concerns what we might term our ongoing self-invention – emphatically not self-discovery – as a project of sustained critique. This process demands recognition of our own irreducible historicity alongside acknowledgment of the historical contingency of those narratives and normative frameworks constructed within discourses that order, regulate, subjugate, and domesticate contemporary experience.

The critical imperative extends beyond the historian's conventional purview. It becomes incumbent upon us to acknowledge our own temporal embeddedness as well as the historical contingency of narratives and norms – including those associated with humanism – which emerge from discourses that systematically order, standardize, regulate, and subjugate present conditions. Such acknowledgment constitutes neither resignation nor relativism, but rather the prerequisite for genuine critical engagement.

The historian's responsibility encompasses ensuring that critical analysis does not terminate at the historical level but extends into contemporary praxis. This involves acknowledging both our own historicity and the temporal specificity of discursive constructions – including humanistic discourse. In this sense, historicization transcends mere academic exercise to become nothing less than a critical project of freedom.

Leonardo O. Moreira*

WILHELM VON HUMBOLDT
Humanism and the 'Cosmopolitanism of Differences' Challenged by the New World

Abstract

Bildung occupies a central place in the humanism that emerges from Wilhelm von Humboldt's anthropological philosophy. It is through *Bildung* that individuals can develop their strength, cultivate their inner freedom, and thus forge their individuality. *Bildung*, as a vector for the realization of human nature, constitutes a universal process that extends to all humanity; it represents the focal point of Humboldt's humanism, which is reflected in his idea of cosmopolitanism and in his works on politics, comparative anthropology, and linguistics. Can Humboldt's cosmopolitanism and humanism, as well as his dialogical thought that gradually develops in these fields of study, achieving a synthesis between the speculative and the empirical, remain unchanged in the face of the radical alterity of the non-European? Can the relativism inherited from Herder withstand the extension of *Bildung* to peoples described as 'savages', as a privileged means of a conversion process? The research hypothesis suggests that an analysis of the *Essai sur les langues du Nouveau Continent* (1812), considered in relation to anthropological and linguistic reflection, can reveal, from within the internal economy of Humboldt's energetic philosophy, the limits of his cosmopolitanism, the paradoxes of his humanism, the ambivalence of his relativism, and the contradictions inherent in his dialogical thought.

Keywords

Wilhelm von Humboldt – Humanism – Cosmopolitanism – New Continent – Conversion

* Université Paris 8.

1. *Introduction*

It has already been observed that Wilhelm von Humboldt's humanism 'appears [...] as the constant pursuit of a balance between the idea, humanity, and the originality of the individual and cultures'.[1] In other words, Humboldt conceives the idea of humanity as the representation of what is essential in human nature, in the Aristotelian sense of a perfected nature. In any case, this represents an 'intensive and ideal greatness',[2] to borrow Humboldt's own terms from his booklet on the eighteenth century. This 'intensive and ideal greatness', although conceived as an unavoidable ontological paradigm, remains inaccessible to any comparative method claiming scientific rigor. Nevertheless, it must be conceived and considered as a referential standard, so that the ends of nature may be achieved. To fully grasp the plural character of humanity, it is necessary, as Humboldt notes, to reflect on 'the possible extension of the character of humanity', since this character 'encompasses everything that arises from human dispositions and tendencies, [and] is not precisely determined anywhere, but may be so everywhere'.[3]

The immutable principles of pure philosophical speculation on humanity in general must, at the risk of falling into arbitrariness, give a predominant place to experience, even to the point of subordinating themselves to it. Nevertheless, this subordination does not imply that there exists only a single intensity of energy constituting the univocal essence of humanity. For humanity manifests itself in time and space through 'diverse forms' which, without exception, undergo continuous phylogenetic change. These forms vary, transform, and develop under the influence of chance. Yet, within this contingent extension, it is reason — unfolding through *Bildung*, and naturally manifesting in each individual, or more precisely through 'the force

1 J. Gandouly, 'Wilhelm von Humboldt ou l'humanisme comme idéal de vie et objet de recherches', *Revue d'Allemagne et des pays de langue allemande*, 25.3 (1993), p. 273.

2 Wilhelm von Humboldt, *Le Dix-huitième siècle*, in *Le Dix-huitième siècle: Plan d'une anthropologie comparée*, trans. by Christophe Losfeld (Lille: Presses Universitaires de Lille, 1995), p. 33.

3 Ibid.

of our will' that proceeds from it — that impels human beings to transcend the indeterminacy of chance in order to achieve the ends determined by reason itself. Humboldt begins with the observation that, even in the Enlightenment, there persists not only a great diversity of individual and collective characters specific to different nations, but also a notable disparity in education and formal processes of self-perfection. Many may believe that each individuality or culture, despite its differences, possesses an ontologically equal status within this anthropological economy.

In an exegetical passage from the third volume of his *Writings* [*Écrits*], Étienne Balibar examines two models of cosmopolitanism, in order to explore, throughout modernity, both the cosmopolitical stakes and the effective conditions of strangeness. The first model, that of Kant, certainly informed later formulations aimed at conceiving cosmopolitanism as the political mediation of human rights, insofar as it contributes to the relativization of the absolute power of the sovereign, while adhering to a liberal orientation subordinated to moral imperatives. The vulnerabilities of this cosmopolitanism become apparent when it is examined in the light of Kantian anthropology: the categorical imperative undergoes notable exceptions, while the very possibility of strangeness is compromised within the horizon of an anthropological scale inherited from Buffon and the notion of perfectibility that runs through the Enlightenment. The second model, developed by Wilhelm and Alexander von Humboldt, takes a different path: it departs from the primacy of an ontological identity founded on resemblance, and is thus configured as a kind of 'cosmopolitanism of differences'.[4]

For Balibar, the Humboldt brothers carried out, between 1799 and 1858, a major reversal in the relationship between the universalist ideal and difference. Eurocentrism, which accompanies this universalism and is intrinsically linked to colonialism, is thus partially contradicted – albeit in a biased manner – since the idea common to the Humboldts rests on the 'equality of cultures as creations of human genius, relative to a certain environment or geographical

4 Étienne Balibar, *Cosmopolitique: Des frontières à l'espèce humaine, Écrits III* (Paris: La Découverte, 2022), p. 84.

habitat'.[5] Yet, the questioning of this global relativism is, according to our hypothesis, constrained by Wilhelm von Humboldt's energetic and dialogical conceptions of linguistic difference, particularly when confronted with cultural difference, as in his *Essay on the Languages of the New World* [*Essai sur les langues du Nouveau Continent*], written in French in 1812 and subsequently published in 1904 (a text generally neglected by defenders of Humboldtian cultural relativism[6]). Balibar nonetheless highlights a possible ambiguity in Humboldt's linguistic studies, because he did not adopt a definitive position regarding the relationship between the different degrees of perfection of languages. Despite this caveat, Balibar argues that 'egalitarianism is the logical tendency of his theory'.[7] It would be pertinent to confront Humboldt's perception of so-called 'savage' peoples with contemporary indigenous thought, which undoubtedly has much to contribute to the debate. However, we will limit our discussion to the exegetical framework. From this perspective, we advance the interpretation that egalitarianism within difference appears, nevertheless, unachievable within Wilhelm von Humboldt's energetic philosophy. Indeed, every differential singularity proves to carry a difference in the intensity of the development and perfection of its own energy.[8] In this way, the 'cosmopolitanism of differences' evoked by Balibar, or the 'comparative study of languages' presented as 'a manifesto of diversity' by Denis Thouard,[9] as well as Humboldt's humanism and dialogical theory, find themselves at

5 Ibid., pp. 89–90.

6 Even if it is, as Jürgen Trabant asserts, one of 'the major texts of Humboldt's philosophy of language'. Jürgen Trabant, 'De l'anthropologie parisienne aux langues du nouveau continent', *Les études philosophiques*, 113.2 (2015), *Wilhelm von Humboldt et la liberté des langues*, ed. by Denis Thouard, pp. 215–38 (p. 223).

7 Balibar, *Cosmopolitique*, p. 90.

8 This explains his predominant interest in peoples 'where human activities are most developed, where they manifest to the fullest what human beings are capable of doing'. Trabant, 'De l'anthropologie', p. 216.

9 Denis Thouard, 'Liberté, individualité, diversité : Humboldt, de la politique aux langues', in *Le libéralisme de Wilhelm von Humboldt: Autour de l'Essai sur les limites de l'action de l'État*, ed. by Olivier Agard and Françoise Lartillot (Paris: L'Harmattan, 2015), pp. 19–33 (p. 32).

an impasse, ultimately reintroducing a universalist telos tinged with Eurocentrism, which is also capable of legitimizing the civilizing enterprise, in a marriage of education and conversion.

Humboldt's perspective, in the *Essay on the Languages of the New World*, on the Amerindian peoples, whom he describes as 'savage', as well as on their languages, also termed 'savage', is not separated from the practical and moral perspective of his humanism (following Kant), nor from the cultural relativism transmitted to him by Herder. Despite the legacy of this relativism, Humboldt considers, in the name of his humanism, the most effective means to cultivate and shape anthropological difference, so that the Jesuits might educate, convert, and civilize the Amerindian peoples. In other words, Humboldt enunciates the formula that he would develop a decade later in his treatise *The Task of the Historian* [*La Tâche de l'historien*, 1822], according to which 'to understand always consists in the application of a preexisting generality to a new particular being'.[10] This formula is explicitly transposed to anthropological difference in the *Essay on the Languages of the New World*. Such a finding raises questions about the ways in which his 'cosmopolitanism of differences' and his humanism become mired in ambiguity, while his dialogical thought is reduced to a monological logic.

2. *The Interest in the Languages of the New Continent*

Humboldt holds that, owing to its itinerant ways of life, its territorial dynamics, and its geographical determinants, America displays a profusion of languages unknown in the other 'regions of the Old Continent'.[11] This great quantity and diversity of languages would persist despite their ephemeral nature: while some languages die out, others are continually born and transformed. These nascent and evolving languages, in addition to offering a vast diversity of forms, may be regarded as a living laboratory for the universal study of languages. Although far removed from a primitive state,

10 Cited in Jürgen Trabant, *Traditions de Humboldt* (Paris: Éditions de la Maison des Sciences de l'Homme, 1999), p. 52.

11 Ibid.

the language of the Amerindians would nonetheless be classified as existing in a kind of 'childhood' of speech, when compared with languages that have attained a higher degree of complexity. This persistence at the threshold of language, despite the fractures of time, bears witness to the ability of Indigenous communities to preserve, beyond appearances, a living bond with their origins. It suggests that Amerindian identity, far from dissolving in history, is constantly reinventing itself, integrating new realities without renouncing the foundations that define it. Thus, this *great individualising figure* becomes a vector of cultural resilience, providing a frame of reference for present and future generations, while enriching the global understanding of identity processes in a world in perpetual transformation.[12] It also helps illuminate certain phenomena that, as Humboldt notes, shed light upon 'the most important questions of the primitive history of nations' and upon 'the mystery of the formation of languages'.[13] As we read in the *Essay on the Languages of the New World*, 'the study of the Indian languages, insofar as we can undertake it from the data at our disposal, offers us another advantage, equally great and important – namely, that of broadening our conceptions of languages in general'.[14] To carry out this part of his comparative study of languages, Humboldt envisaged a taxonomic approach aimed at gathering, organising, and methodically classifying all existing material concerning the languages of the New Continent.[15]

3. *To Study, Convert, and Civilise: The Role of Religion*

Our analysis does not, however, dwell on the scientific and philosophical interest aroused by linguistic difference taken in isolation, but rather on the teleological ideology of the civilising

12 Wilhelm von Humboldt, *Introduction à l'œuvre sur le kavi, et autres essais*, trans. by Pierre Caussat (Paris: Éditions du Seuil, 1974), p. 146.

13 Wilhelm von Humboldt, *Essai sur les langues du Nouveau Continent*, in *Wilhelm von Humboldts Werke 1799–1818* (Berlin: B. Behr's, 1904), p. 303.

14 Ibid., p. 306.

15 Ibid., p. 307.

mission that insinuates itself within it. There is no doubt that Humboldt was careful not to present himself as a theorist of a process of civilisation marked by the brutality of the colonisers. His cosmopolitanism could never have risen – as it does, for instance, in Balibar – to the level of a 'cosmopolitanism of differences' by following such paths of violence. Even the *Essay on the Languages of the New World* illustrates his cultural relativism by criticising the acculturation imposed by the Jesuits on the Amerindian nations. By imposing the Christian religion and forcing the peoples into conversion without adequate prior education, the Jesuits, Humboldt writes, had initially set out to 'uproot, along with their former practices, everything connected to their traditions and to the national memories [of the Amerindians], and thus to change completely their way of thinking and feeling'.[16]

Although permeated by this relativism, this critique intertwines with a normative vision that duplicates the anthropological-political normativity previously advocated, namely the pedagogical task of his comparative anthropology and the more restrictive task of the state acting objectively upon the individuality of its members.[17] What Humboldt critiques is not conversion itself, nor the civilising mission undertaken by the Jesuits, considered in isolation, but rather the method then employed and the highly uncertain quality of its results, both for the study of languages and for the conversion of Indigenous peoples.[18] Indeed, Humboldt projects onto the Indigenous peoples the educational plan he idealises for Europe: the educational approach of his humanism aspires to extend to all of humanity, which shares a single ideal of moral improvement. This role of religion as a universal vector was also emphasised in *The Limits of State Action* [*Essai sur les limites de l'action de l'État*], written in 1792 and published in 1851:

16 Ibid., p. 304.

17 As Humboldt writes, 'The state seizes the individual having reached his development and maturity; it directs his entire life by directing his manner of acting and his way of thinking'. Wilhelm von Humboldt, *Essai sur les limites de l'action de l'État*, trans. by Henri Chrétien (Paris: Germer Baillière, 1867), p. 86.

18 Humboldt, *Essai sur les langues du Nouveau Continent*, p. 304.

> When the Christian religion, instead of the deities specific to each nation, came to teach the existence of a universal deity common to all men, one of the most disastrous walls separating the different human races from one another collapsed; it thus laid the foundations for all true virtue, all genuine progress, and all authentic union within humanity […].[19]

In his *Plan for a Comparative Anthropology* [*Projet d'une anthropologie comparée*], Humboldt highlights the practical function of religion, namely to teach truth, which is both objective and universal. It is therefore not religion as such that constitutes truth, but it possesses the power to teach it, to transmit it objectively. The truth Humboldt wished the missionaries to convey to the Indigenous peoples of the Americas was not precisely the existence of God, but rather universal morality, which he definitively separates from the dogmas of any particular religion. As Jean Quillien would affirm, if Humboldt acknowledges a God, it is indeed within humanity that he discovers Him. This does not, however, mean that humanity constitutes the content of God. In fact, this content is of relatively little significance in Humboldt's eyes, for he dissociates religious feeling and its moral force from any notion of substance. The utility of religion for practical learning is revealed much more in its form: faith, conviction.

Humboldt's humanist demand for the comparative study of languages is aimed primarily at those who manipulate divinity. With religion, the goal is to mobilise one form of energy in support of the development of another. This is therefore not a paradox between subjectivity and objectivity. For Humboldt, religion remains resolutely subjective; its great value, like that of God, lies in its capacity to function as a force propelling processes of human perfection, even on a global scale. When applied to the Indigenous peoples of the New World, this process reveals, to borrow the terms of Lévi-Strauss in *Race and History*, 'the paradox of cultural

19 Humboldt, *Essai sur les limites de l'action de l'État*, p. 86.

relativism', a kind of 'false evolutionism', since 'diversity is only apparent'.[20]

4. *The Erasure of Linguistic and Anthropological Differences*

Taking into account a 'cosmopolitanism of differences' or Humboldt's humanism, some might argue that, in the encounter envisioned with Indigenous peoples, the strength and energy of European civilization were meant to organize the processes of individuation so that they might reach the universal ends of humanity, and therefore of nature. In his book *The Traditions of Humboldt* [*Les Traditions de Humboldt*], Jürgen Trabant offers a comparative analysis of the parallels drawn by Vico and Humboldt between the Amerindians and Antiquity. He then distinguishes two completely opposed linguistic, cultural, and political processes: first, in Antiquity, where the principle of synthesis prevails – an amalgamation of two entities which, within an intrinsic union, each preserve their own cultural integrity. Then, in contrast to this principle, the New Continent is perceived as the anti-Greece, since it embodies the process of incorporation. Rather than achieving the union of two distinct entities, this process implies 'the destruction of one within and by the other'. The example evoked is that of the Aztecs, for whom this principle manifests itself on both the political and linguistic planes.

Trabant also notes the similarity between this latter principle and the *modus operandi* of the Spanish colonizers, who not only incorporated Indigenous peoples and their cultures but also violently exterminated them. In the note to the *Essay on the Languages of the New Continent*, where Humboldt sets out his reflections on the proper method for civilizing and converting the Indigenous peoples of the Americas, he evokes a 'charitable love that should unite all men', as in a marriage.[21] This forms part of the universal ideal of his

20 Claude Lévi-Strauss, *Race et Histoire* (Paris: Éditions Gonthier/UNESCO, 1961), pp. 22–4.

21 'The great purpose of civilizing and converting the 'savage' nations – which, even considered solely from the perspective of the charitable love

humanism, or, more precisely, the threshold of his cosmopolitanism. A doubt nevertheless remains: what, in fact, are the effects of such a 'marriage' between Europeans and Amerindians within the internal economy of Humboldt's energetic philosophy? I believe that historians have failed to grasp the essence of the question.

On the one hand, at least if we consider countries such as Germany, England, and France, we observe a dual development of energies. Civilization brings forth a high level of anthropological energy, synonymous with a high degree of humanity. This elevated level of energy is inseparable from the development of communicative energy, namely language, which itself is inseparable from thought. On the other hand, the American nations, lacking a well-structured language, have not, according to Humboldt, further developed their cognitive power, that is, thought. If each language corresponds to a worldview, and if, as José Voss observes, 'man can access this purely objective domain only through the subjective path [...] of cognition and volition indicated to him by his mother tongue',[22] this means that the perceived world is circumscribed within the limits of their language. The horizontality of relativity between different worldviews remains always subordinated to the verticality of each in relation to an ideality of worldview, in which energies articulate forces to a high degree of intellectual individuation.

Every human being has been endowed by nature with a 'primitive force, the original Self', and it is upon this force that 'the freedom of man rests'. Any increase in a person's energy and strength, as well as in their language, goes hand in hand with the development of this Self, this primitive force inherent to each individual. Although Humboldt opposes a materialist view according to which man 'was merely the product of a pure mechanism', defending the role of each individual's primitive force in their formation as a person, Indigenous peoples would be disadvantaged in their relation to the outside world, whether due to the isolation to which they were

---

that should unite all men – would be of such immense importance [...]'. Humboldt, *Essai sur les langues du Nouveau Continent*, p. 304.

22 José Voss, *Le langage comme force selon Wilhelm von Humboldt* (Paris: Connaissances & Savoirs, 2017), p. 264.

subjected, the natural conditions that constrained them, or the absence of intergenerational transmission. All these characteristics stand in equal opposition to Humboldt's dialogical thought. As José Voss explains – using Heidegger's terminology – for Humboldt, 'the human condition is dialogical', arguing that 'authentic dialogue involves what Heidegger designates as being-with-others [...]'.[23]

Isolation and dispersion therefore end up also limiting the dialogical dimension in this experience of anthropological difference. Consequently, it would not be illusory to imagine the asphyxiation of their primitive forces, and thus of their 'original Self'. Since, as we read in *Plan for a Comparative Anthropology*, man can access true individuality only by first attaining a substantial degree of culture. It is the intellectual faculty itself – the judgment proper to otherness – that proves formless, being devoid of singular concepts. It is in form that we can grasp the difference of a language, and, as Humboldt states, this form is 'closely linked to the spiritual dispositions of nations and to the energy that elevates them at the moment of their emergence or innovative impulse'.[24] The alleged limits of language, and consequently of thought, restrict difference both to the incapacity to continue being and to the power to yield to another 'higher' language-form. One could discern here the morphological aspect of his energetic philosophy. Form, energy, and individual freedom occupy a central place in Humboldt's anthropological thought. The exercise of individual freedom is paramount, as it constitutes the very basis for the perfection of liberty. There would be neither a civilizing project nor a method for proper conversion if the Indigenous peoples of the Americas enjoyed greater individual freedom. When, in *The Limits of State Action*, Humboldt considers collective action necessary for the realization of this freedom, he seeks to demonstrate that the proper dynamic of forces lies on the side of civilization, in opposition to the way of life of 'savage peoples': 'It is enough to compare', says Humboldt, 'the well-being of civilized and enlightened societies with the misery of

23 Ibid., p. 292.

24 Humboldt, *Introduction à l'œuvre sur le kavi*, p. 190.

barbarous and savage peoples, to be convinced of the truth of this proposition'.[25]

By setting out the terms in this way, we are faced with a marriage, an unbalanced encounter, involving unequal forces and energies. Not only acquired forces, but also primitive ones. If, as Humboldt notes in his short treatise *On the Eighteenth Century* [*Sur le XVIIIe siècle*], 'nature rebels against the senseless attempt to eradicate the given character in order to adopt one that is completely foreign to it', since 'this reduces it to nothing', might nature possess the same power with regard to the Indigenous nations of the Americas? It must be considered that nature holds this power only when this given character follows the designs of reason, for it cannot contradict that natural path. Regardless of cultural differences and the particularities of each individual force, we should remember that all these forces together constitute the idea of humanity and its elusive reality. However, we must pay attention to humanist teleology: all humanity must follow the same model and reach the same end – that of intellectual and moral perfection. In this sense, the encounter with anthropological difference, the civilizing project, and the work of conversion would not oppose the primitive force of individualities, but rather orient them toward the right paths to achieve their ends.

Let us not lose sight of the fact that Humboldt anticipates, in his *Essay on the Languages of the New Continent*, that all languages in regions where Europeans penetrate will gradually disappear. However, this is not an isolated rule, nor merely the product of prejudice against otherness. This law applies invariably to every asymmetrical relationship between two languages whose respective forces are in imbalance. Such is the case in this reflection on the conversion of the Amerindians, which pursues the ideal of civilizing them through a progressive purification of their religion, until they effortlessly recognize the true source of truth in the Christian religion.[26] To carry out such a design, the missionaries must reflect

25 Humboldt, *Essai sur les limites de l'action de l'État*, p. 104.

26 'The natural course, then, would be gradually to purify the religion of the savages [...] to make them see, little by little, that divine goodness has sown sparks of truth everywhere, but that there exists one religion where its source flows abundantly [...] with no other interest than that of truth

on communication, that is, on the means by which they will transmit intellectual and moral teachings.

According to Humboldt, languages are not meant to merge to form a united couple; they fully express their power only when they remain faithful to their primitive roots. Thus, when they tend to come closer, it is always the weaker language that must give way. If, within this logic – where a proximity to the Darwinian idea of natural selection can be discerned – 'savage languages' must disappear, it is not only the language that yields, but thought itself, since the existence and development of one are inseparable from the existence and perfectibility of the other. They are energies that are organically dependent. A simple reading of the reports and letters of the missionaries suffices to reveal, as Humboldt asserts, how most of the indigenous languages suffered and declined under the weight of European domination. Upon closer analysis, it becomes clear that by civilising the Amerindians, by leading them to the Christian religion and imposing a new language upon them, it was not only the 'supposed errors' of their religion that were to be 'purified', but their very individuality that was diverted from its contingent trajectory.

To indicate the right path and to provide the means of attaining the common ends of humanity would still, according to Humboldt, constitute an act of charity. He considers that even when left to themselves, the American nations are exposed to many trials and upheavals, and that the survival of their languages is often *ephemeral and unstable*.[27] If this energy that is language appears fleeting and mobile, the same can be said of thought, so that their primitive force remains in a state of ontological precariousness. Indeed, Humboldt writes,

> the original properties of very primitive peoples are, for the most part, either external or accidental, and therefore insignificant. Sometimes, it is simply an insufficient development that serves as their distinguishing feature.[28]

---

and their happiness […]'. Humboldt, *Essai sur les langues du Nouveau Continent*, p. 304.

27 Humboldt, *Plan d'une anthropologie comparée*, p. 187.

28 Ibid., p. 181.

It is therefore the entire dialogical dimension that collapses in its effective relation to otherness, revealing the limits of his 'cosmopolitanism of differences', as well as of his humanism. Despite all that *Bildung* should offer in encountering human diversities – that is, the capacity to welcome another religion or a different way of life – and in the face of the inevitable disappearance of languages deemed less developed, what we witness is indeed the dissolution of the dialogical framework into a monologue.

The terms of the paradox of a 'cosmopolitanism of differences' and Humboldt's humanism become clearer in the light of the energetic logic of his comparative anthropology and his conception of language. However, it remains impossible for us to fully grasp them without opening ourselves to the perspective of otherness, for which the civilizing aim emerging from reflections on the conversion and education of indigenous peoples is not devoid of significance. To conclude, we reproduce an excerpt from the *Outline of a Comparative Anthropology* [*Plan d'une anthropologie comparée*], which clearly illustrates the inherent ambiguity of Humboldt's relativism, as well as the paradoxes and limits of his humanism and its dialogical conception.

> Comparative anthropology therefore pursues a dual objective and generates a dual interest. It facilitates the understanding of human characteristics and allows for the establishment of the philosophical criteria necessary to judge them, to estimate their future development, and, ultimately, to assess their capacity to act with others toward a common goal. It proves fundamental for the businessman who seeks to use and dominate men, yet it is equally useful to the educator and the philosopher, who continually strive to educate and perfect them.[29]

Translated by Caio Francisco Azevedo

29 Ibid., p. 168.

Anna Pia Ruoppo*

INVENTING THE RENAISSANCE
Jacob Burckhardt and the discovery of the individual

Abstract

After the barbarism of the Second World War, many intellectuals questioned the possibility of restoring meaning to Humanism. But what is meant by this fundamental concept? This text examines the genealogy of Humanism, first retracing its appearance in public discourse, then its codification in Jacob Burckhardt's text, *The Civilization of the Renaissance in Italy*. If Humanism arose from the discovery of the individual, and if the individual is the individual described by Pico della Mirandola's *Oration on the Dignity of Man*, then rejecting or accepting Humanism means taking a stance on modernity and its understanding of the subject and its autonomy, as clearly demonstrated in the debate between Sartre and Heidegger.

Keywords

Modernity – Humanismus – post-World War II – Sartre – Heidegger

1. *Renaissance and modernity*

The title of this paper is taken from a chapter in the text *Europa 1959*, published in 2023 by Arthur Haberman, with the subtitle 'In the Ebb and Flow of Modernity'.[1] In this book the author considers 1859

* Università degli Studi di Napoli Federico II.

1 Arthur Haberman, *Europa 1959: In the Ebb and Flow of Modernity* (Berlin/Boston: De Gruyter, 2023).

as a pivotal year for the birth of the category of modernity: this is 'a *Wunderjahr*, a wonder-year, in the cultural history of the European continent'[2] in which 'important thinkers and public intellectuals encountered modernity and reflected on it'.[3] In 1859, Baudelaire was 'writing the poetry and criticism of the new urban cultural and social world which would make him described by a number of historians as the first modern'.[4] In the same year, 'in the east, Ivan Turgenev with *On the Eve* [began] reflections about Russia and modernity which would result in his next novel, set in 1859, *Fathers and Sons*'.[5] In England, two authors in exile, Karl Marx and Giuseppe Mazzini, were working respectively on *The Capital* and *The Duties of Man*, while John Stuart Mill published *On Liberty* in February 1959 (laying the foundations for modern democratic theories and an understanding of the relationship between freedom and authority). In November of the same year, Charles Darwin published *The Origin of Species.* In this context, as the European map was beginning to change considerably, Jacob Burckhardt, according to Haberman, in his book *Die Kultur der Renaissance in Italien,* published in 1960, 'invented the Renaissance to better understand his own time'[6] and 'totally transformed how we think about Italy in the Renaissance and the meaning of the Renaissance itself'.[7]

The way in which Burckhardt revolutionises the conception of history would deserve consideration in its own right, but this is not the topic I want to dwell on in my article. My intention here is rather to examine Burckhardt's text starting with the problem we have set ourselves, i.e. to provide a conceptual historical genealogy of Humanism and of the Renaissance in order to better understand today; that is, to better understand what we are doing here and now. In brief, where does that image of Renaissance and humanism, the subject of our research and discourse, come from, regardless of whether we are supporters or critics of it? The specific object of

2 Ibid., p. 9.
3 Ibid.
4 Ibid., p. 8.
5 Ibid., pp. 8–9.
6 Ibid., p. 8.
7 Ibid., p. 61.

my current research and my contribution to the project 'Paradoxical Humanism and Environmentalism' is an analysis of the post-World War II discussions focussing on the question of whether it is possible to restore meaning to humanism. The point of view that guides intellectual discussions around this question, which is closely linked with the question of what Europe was and can still become after the war, emerges in the essay of the French philosopher Maurice Merleau-Ponty in the first issue of *Les Temps modernes*, the magazine he founded immediately after the war together with Jean-Paul Sartre. Here the philosopher affirms: 'We were not wrong, in 1939, to want liberty, truth, happiness, and *humanism*. [...] It is a question not to give up our values of 1939 but to realize them'.[8] The first to intervene in the public debate was Jean-Paul Sartre who in October 1945 gave a lecture entitled *L'existentialisme est un humanisme*.[9] The echo of this lecture also reached the German philosopher Martin Heidegger who implicitly responded to Sartre with his never delivered *Brief über den Humanismus*.[10] In 1949 Merleau-Ponty himself published the text *Humanisme et terreur*,[11] and Karl Jaspers, accepting the organisers' invitation to close the IV *rencontres internationales de Genéve* with a programmatic speech, gave the lecture *Über Bedingungen und Möglichkeiten eines neuen Humanismus*.[12] But what are the four intellectuals talking about? Does the humanism they speak of refer to that movement of culture that established a new vision of the world and man between the 15th and 16th centuries? Or is it a pedagogical ideal, the *Bildung*, or, more accurately, is it the expression of a certain conception of man? Or something else? A clarification of the different ways to say *Humanismus* will allow us to better understand the post-war discussions.

8 Maurice Merleau-Ponty, *Sense and Non-Sense* (Evanston: Northwestern University Press, 1964), p. 152.

9 Jean Paul, Sartre, *L'existentialisme est un humanisme* (Paris: Nagel, 1946).

10 Martin Heidegger, *Brief über den Humanismus* (Frankfurt am Main: Vittorio Klostermann, 1949).

11 Maurice Merleau-Ponty, *Humanisme et terreur* (Paris: Gallimard, 1948).

12 Karl Jaspers, *Über Bedingungen und Möglichkeiten eines neuen Humanismus* (Stuttgart: Reclam, 1951).

2. *A new term appears in scientific discourse*

Indeed, the term humanism appears for the first time in the *Reisetagesbuch*[13] of the theologian Johann Friedrich Abegg, who on 25 April 1798 set out from Boxberg, a small village near Würzburg, to join his brother in Königsberg. During the five-month journey Abegg meets writers and philosophers of German classicism and idealism (including Kant). But it is in a dialogue with the jurist Johann Georg Scheffner, from Königsberg, in fact, also a friend of Kant, that the term *Humanismus* appears in a sense which strikes me as not entirely positive. The two spend a pleasant morning speaking about recent political events, in particular about the effects of the French Revolution on Prussia and Germany, when the lawyer from Königsberg expresses his criticism of all those professional intellectuals who think only 'through the book, in the book and for the book',[14] whereas it is in business that men are formed, where they experience real life. It is in this context that Scheffner enquires about the University of Heidelberg (where Abegg teaches practical theology) and asks whether Humanism is also taught there. The theologian's ambiguous and over-interpreted answer is: 'No, they do not appreciate the fine Wines and their Spiritus (Spiritus) there'. This example informs us of the way in which the term *Humanismus* was used in everyday language, even before it made its appearance in public discourse when, in 1808, Friedrich Immanuel Niethammer as central commissioner for education and member of the General Consistory of Bavaria, published the booklet *Der Streit zwischen Philanthropinismus und Humanismus in der Bildungstheorie unserer Zeit*,[15] with a view to reforming the school system, which he later managed to realize. 'Philanthropism' and 'Humanism' represent two opposing pedagogical conceptions. Philanthropism has the goal of 'the formation of man with a view to his future action in the world',

---

13 Johann Friedrich Abegg, *Reisetagesbuch von 1798* (Frankfurt am Main: Insel, 1987).

14 Ivi, p. 235 (my translation).

15 Friedrich Immanuel Niethammer, *Der Streit des Philanthropinismus und Humanismus in der Bildungstheorie unserer Zeit* (Jena: Frommann, 1808).

the subjects of classes must cover all areas of modern knowledge, 'in particular physical and technical knowledge' and must not be limited to 'the sphere of antiquity'. On the contrary, *Humanismus* aims at 'the general education of man', which must precede 'the preparation of the learner for his professional life'. This training must take place through the study of those texts to which is attributed 'authentic classicism in all the ways of the representation of the true, the good and the beautiful'. The pedagogical model to which the label *Humanismus* is attributed is that which would be later be defined as *Neohumanism* of Humboltian imprint.[16] In Niethammer's text there is no explicit reference to Renaissance humanism; however, the historian Ulrich Muhlack, in his volume *Humanismus und Renaissance*, defines his *Humanismus* as 'a phenomenon of Renaissance', in that it takes its name from the *studia humanitatis* on which the curriculum is to be centred to form the learner as a man.

The term *Humanismus* is used about thirty years later as a historiographical category both in Karl Hagen's text *Deutschlands literarische und religiöse Verhältnisse im Reformationszeitalter*[17] of 1841 and in Georg Voigt's 1859 text *Die Wiederbelebung des classischen Alterthums oder das erste Jahrhundert des Humanismus*,[18] the former focusing mainly on German history, the latter on Italian history. The term Renaissance also appears in philosophical discourse about a century before becoming a technical term to determine an entire epoch, as in Burckhardt's text. An example is D'Alembert's use of it in his *Le Discourse préliminaire de l'Encyclopédie*,[19] in which he refers to the Renaissance of the arts, literature and philosophy in the period following the Middle Ages. It is the French historian Jules Michelet who uses the term to connote a historical epoch characterised by the 'discovery of the world

16 See Ulrich Muhlack, *Renaissance und Humanismus. Enzyklopaedie deutscher Geschichte*, bd. 93 (Berlin/Boston: De Gruyter, 2017), p. 152.

17 Karl Hagens, *Deutschland literarische und religiöse Verthaeltnisse im Reformationszeithalter* (Bern, 1841).

18 Georg Voigt, *Die Wiederbelebung des classischen Alterthums oder das erste Jahrhudert des Humanismus* (Berlin: Reimer, 1959).

19 Jean Le Rond, D'Alambert, *Le Discourse préliminaire de l'Encyclopédie* (1751) (Trois Rivières: Les Échos du Maquis, 2011).

and man',[20] a discovery that is, however, attributed to Columbus, Copernicus, and Galileo and not to the Italian humanists.

3. *Burckhardt and the discovery of the individual*

However, it is Jacob Burckhardt's text, *Die Kultur der Renaissance in Italien*[21] published in 1860, that completely transforms the way in which we think of the Renaissance. In the introduction, Burckhardt defines his work as an 'essay', *ein Versuch*, an attempt, as 'no one is more conscious than the writer with what limited means and strength he has addressed himself to a task so arduous'.[22] The outlines of a given civilization present a different picture to each eye, and this is all the truer when a civilisation continues to influence our own. The author says he is 'content if a patient hearing be granted [him], and if [his] book be taken and judged as a whole':[23] 'It is the most serious difficulty of the history of civilization that a great intellectual process must be broken up into single, and often into what seem to be arbitrary categories, in order to be in any way intelligible'.[24] But what is his image of the Renaissance and dare I say also of humanism, given that Burckhardt dedicates an entire chapter of his text to the revival of antiquity (*Wiederentdeckung der Antike*)?

The fundamental trait of this epoch consists in the role, which in the political situation of Italy, characterised by the lack of a unitary state, played by the individual. According to the Swiss historian, for Italy, 'the existence of the ecclesiastical State, and the conditions under which alone it could continue, were a permanent obstacle to national unity, an obstacle whose removal seemed hopeless'.[25]

20 Jules Michelet, *Historie de France. Le Renaissance*, vol. 7 (Paris: Hachette Livre, 1855).

21 Jacob Burckhardt, *Die Kultur der Renaissance in Italien* (Basel: Schweighauser'schen Verlagsbuchhandlung, 1860); English trans. by S. G. C. Mibbleton, *The Civilization of Renaissance in Italy* (Vienna/Oxford: The Phaidon Press/Oxford University Press, 1878).

22 Ibid., p. 5.

23 Ibid., p. 12.

24 Ibid.

25 Ibid., p. 69.

Burckhardt identifies precisely 'in the character of these states, whether republics or despotic regimes, lies, not the only, but the chief reason for the early development of the Italian. To this is due that he was the first-born among the sons of the modern Europe'.[26] The historian writes:

> In the Middle Ages man was conscious of himself only as a member of a race, people, party, family, or corporation – only through some general category. In Italy this veil first melted into air; an *objective* treatment and consideration of the state and of all the things of this world became possible. The *subjective* side at the same time asserted itself with corresponding emphasis; man became a spiritual *individual* and recognized himself as such.[27]

It is the development of the individual that offers the key to understanding the Renaissance as an organic whole.

It is in the third chapter, dedicated to the *Revival of Antiquity*, that the author coins the term Renaissance as a technical term intended to define an entire era. At the same time, the place that the author attributes to humanism, initially understood as *studia humanitatis*, emerges. After addressing the state as a work of art and dwelling on the development of individuality, Burckhardt states:

> Now that this point in our historical view of Italian civilization has been reached, it is time to speak of the influence of antiquity, the 'new birth' – in German, the term is *Wiedergeburt*, but the Italian, translator already uses the word 'Rinascimento' – which has been one-sidedly chosen as the name to sum up the whole period.[28]

Burckhardt emphasises that there is no doubt that the events described so far – the formation of the state as a work of art, the development of the individual –

> would have sufficed, apart from antiquity, to upturn and to mature the national mind and most of the intellectual tendencies which yet remain

26 Ibid., p. 70.
27 Ibid.
28 Ibid.

> to be noticed would be conceivable without it. But both what has gone before and what we have still to discuss are coloured in a thousand ways by the influence of the ancient world; and though the essence of the phenomena might still have been the same without the classical revival, it is only with and through this revival that they are actually manifested to us.[29]

It is here, through a shift in the use of the term from new birth/ Wiedergeburt/Rinascita) to Renaissance (in Italian translated as Rinascimento) that the technical term of Renaissance is coined to indicate an entire epoch. Burckhardt in fact continues:

> The Renaissance – but here the term is no longer Widergeburt/ rinascita, but precisely Renaissance/in Italian, Rinascimento, in the masculine form – would not have been the process of the world-wide significance which it is, if its elements could be so easily separated from one another.[30]

However, Burckhardt writes – and this thesis is fundamental to understand his image of the Renaissance and of humanism – 'we must insist upon it, as one of the chief propositions of this book, that it was not the revival of antiquity alone, but its union with genius of the Italian people, which achieved the conquest of the western world. The amount of independence which the national spirit maintained in this union varied according to circumstances. 'The "Renaissance" would not have been the supreme world necessity that it was, if this impact of the ancient world could so easily have been disregarded'.[31] However, Burckhardt continues – and this thesis is also fundamental to understanding the image of the Renaissance and humanism that he intends to convey – 'we must establish as of now, as an essential point, of our book, that not the impact of the ancient world alone, but it and the spirit of the Italian people, already present, interpenetrated together, had the force to drag the entire western world along with it'.[32] In Italy, the ease with which Latin could be understood, along

---

29 Ibid., p. 89.
30 Ibid.
31 Ibid., p. 161.
32 Ibid.

with the fact that numerous monuments and documents had survived and could be perused by anybody, facilitated the return to the past, and it was the effective participation in all that is ancient, not only by the learned, but by the entire people, that produced 'the modern Italian spirit, which was destined to serve as the model and ideal for the whole western world'.[33]

This interpenetration between the Italian spirit and the rediscovery of antiquity explains both the birth of the new humanist culture and its decline. For Burckhard, in fact, humanism entered a crisis when the vital contact with the past was lost and antiquity was conceived 'dogmatically', as a 'model for all thought and action'[34] and when people 'got used to asking antiquity alone for the solution to any problem', reducing literature to a mere display of quotations. The invention of the printing press and the consequent severing of the link between humanists and the people, led to the decline of humanism and the open and general discrediting of its representatives who were considered a caste.

There is, however, another aspect we must consider to understand how one of the great humanist narrations begins in Burckhard's text. In chapter four, the historian points out how 'freed from the countless bonds which elsewhere in Europe checked progress, having reached a high degree of individual development and been schooled by the teachings of antiquity, the Italian mind [now] turned to the discovery of the outwards universe, and to the representation of it in speech and in form'.[35]

In the middle of the universe there is the human being and 'men and mankind were thoroughly and profoundly understood'.[36] This acquisition of a serious and perfect knowledge of man and humanity in their profound essence does not remain on a theoretical and abstract level, but became a fact in the Renaissance for the first time. According to Burckhardt, 'this one single result of the Renaissance is enough to fill us with everlasting thankfulness'.[37] It

33 Ibid., p. 90.
34 Ibid., p. 140.
35 Ibid., p. 146.
36 Ibid., p. 185.
37 Ibid.

is in this context that the historian quotes a long passage from Pico della Mirandola's *Discourse on the Dignity of Man*, thus opening up the tradition that saw in this text the manifesto of the humanist-Renaissance conception of man and inaugurating the great humanist narration. In Burckhardt words:

> The loftiest conceptions on this subject were uttered by Pico della Mirandola in his speech on the dignity of man, which may justly be called one of the noblest bequests of the great age. God, he tell us, made man at the close of the creation, to know the laws of the universe, to love its beauty, to admire its greatness. He bound him to no fixed place, to no prescribed form of work, and by no iron necessity, but gave him freedom to will and to move. 'I have set thee', says the Creator to Adam, 'in the midst of the world, that you mayst the more easily behold and see all that is therein. I created thee a being neither heavenly nor earthly, neither mortal nor immortal only, that thou mightest be free to shape and to overcome thyself. Thou mayst sink to a beast and be born anew to the divine likeness. The brutes bring from their mother's body what they will carry with them as long at they live; the higher sprits are from the beginning, or soon after, what they will be for ever. To thee alone is given in thee the germs of universal life.[38]

The invention of the Renaissance also coincides with the invention of humanism as that doctrine that identifies man's dignity in his freedom. Highlighting this aspect, the Swiss historian closes his narration thus:

> While the men of the Middle Ages look on the world as a vale of tears, which Pope and Emperor are set to guard against the coming of Antichrist; while the fatalist of the Renaissance oscillates between seasons of overflowing energy and seasons of superstition or of stupid resignation, here, in this circle of chosen spirits, the doctrine is upheld that the visible world was created by God in love, that it is the copy of a pattern pre-existing in Him, and that He will ever remain its eternal mover and restorer. The soul of man can be recognizing God draw Him into its boundaries, but also by love to Him itself expand into the Infinite – and this is blessedness on earth. Echoes of medieval mysticism here flow into one current with Platonic doctrines, and with

38 Ibid.

> a characteristically modern spirit. One of the most precious fruits of the knowledge of the world and of man here comes to maturity, on whose account alone the Italian Renaissance must be called the leader of the modern ages.[39]

So what is the image of the Renaissance that Burckhardt returns to us?

The pivot around which this new civilisation revolves is the extraordinary development of the individual, whose spiritual rebirth would have been possible even without reference to Antiquity. It is the renewed focus on the latter, however, that allows this renewal to be given an external form.

Humanism, understood as *studia humanitatis* and philology, retained its strength as long as it was rooted in concrete life and gave voice to the vitality of the individual, while it entered a crisis, when with the discovery of printing, that link between the individual (humanist) and the people was broken. But there is another reason that determines the crisis of humanism: it is the transformation of Antiquity from a way of giving voice to renewal, to an ideal towards which all thought and action is oriented.

However, it is on that assertion that, during the Renaissance, a perfect understanding of man and humanity was achieved in practice, that the idea of humanism, not as *studia humanitatis* or *Bildung,* but as a reflection on the nature and essence of man, is based. The fact that Burckhardt himself, when acknowledging the Renaissance's achievement of a perfect understanding of man and humanity in practice, refers to Pico's oration, has contributed to identifying that renaissance humanism with the codification of the essence of man, which identifies dignity in freedom. This is where the great humanist narration starts. Now back to our initial question: does this genealogical path help us shed light on the post-World War II debates? Obviously, we can only hint at the question here, focusing on the distance dialogue between Heidegger and Sartre. It is in this dialogue that the debate on the possibility of rediscovering the meaning of humanism becomes the testing ground for modernity

39 Ibid., p. 392.

and leads to a break that would influence all future reflections on the essence of man as existence and on the way of understanding knowledge and philosophical reflection.

4. *The struggle around humanism: Sartre and Heidegger*

Turning from his earlier critique of the encyclopaedic humanism impersonated by the Autodidact in *La Nausée*, Sartre calls his existentialism a humanism, because it reminds man that he is the sole legislator of himself, and that his dignity lies in his freedom.

In this identification of man's dignity in his freedom, it is possible to read an implicit reference to Pico della Mirandola's oration *De hominis dignitate*. There are no explicit references or quotations from the Renaissance philosopher in Sartre's texts, but the fact that a first French translation, although partial, had already been written in 1578 (the first in modern languages), and that in the 20th century the *Discourse* had become established as the most famous of Pico della Mirandola's texts, suggests that the philosopher was aware of the conception of man codified in it.[40] Robert Klein, in a text that has become a classic for Renaissance studies, *La forme et l'intelligible*, also suggests a 'connection' between existentialism and Pico's thinking when he affirms: 'No one before Sartre had had such a strong sense of the total opposition between the being of nature and the soul seen as knowledge, freedom, and the possibility of autonomy as Pico did'. For this reason, according to Klein, 'Pico, in his own way, states unequivocally that of all beings, man is the only one in whom existence precedes essence'.[41]

The most famous part of Mirandola's text – the one Burckhardt also quotes – concerns the speech that the god-architect addresses to Adam after creating the cosmos as a marvellous work. 'Nothing seemed more miraculous than man', 'an amazing creature', but this extraordinary nature had to be understood and identified in

40 On this point, see Raphael Ebgi, 'Introduzione', in Giovanni Pico della Mirandola, *La dignità dell'uomo* (Turin: Giulio Einaudi, 2021), pp. vii–xxxiv (pp. xxxvi–xxxvii).

41 Robert Klein, *La forme et l'intelligible* (Paris: Gallimard, 1970).

the absence of a defined nature. Pico's God does not assign man a fixed place, but places him at the centre of the world, giving him the possibility of establishing his own nature, without being constrained by laws prescribed by God.

God plants within man the seeds of every kind of existence; depending on which ones each person cultivates, they will develop and bear their first fruits within him. A trace of the silent dialogue that Sartre engages in with Pichian oration can be found in that sort of counterpoint to the god-architect's speech to Adam, namely the speech that Jupiter tries to make to Orestes in the play in which Sartre focuses on the discovery of freedom, the drama *Les Mouches*.[42] Written in 1942 and staged the following year, during the Nazi occupation, the play is a reworking of *The Cophorae* by Aeschylus and focuses on the story of Orestes, son of Agamemnon and Clytemnestra, who was taken from Argos at the age of three by Aegisthus, his father's murderer and his mother's lover. After seventeen years, Orestes returns to Argos and, deciding to take charge of his freedom, revenges his father's murderer by committing a double crime. Jupiter, who knows that someone will arrive to announce *the twilight of the gods*, tries to dissuade him from his fatal intentions and, after the murder, makes a last desperate attempt to bring Orestes back into his pre-established order, inviting him to return to his nature and obey his laws. However, he can only be rejected by Orestes, who has now discovered his freedom and can have no other law than his own. By focusing attention on man's freedom and his ability to become what he is, Sartre takes the Renaissance view of man, which began with Burckhard's narrative, to its extreme consequences. Martin Heidegger responds to this question in his *Letter on Humanism*.[43] According to the German philosopher, humanism is always characterised by a contrast between what is human and what is not: the barbarian. From a historical point of view, it is in Rome that we 'encounter the first

42 Jean Paul Sartre, *Les Mouches* (Paris : Gallimard, 1943).

43 Martin Heidegger, *Brief über den Humanismus* (1949). Quoted in the English translation (New York, 1977).

humanism',[44] and the Renaissance of the 14th and 15th centuries in Italy is nothing more than '*a renascentia romanitatis*'. The specific characteristic of 'humanism historically understood' is the revival of antiquity. However, according to Heidegger – and here the reference to the narration started by Burckhard seems to be implied – 'if one understands humanism in general as a concern that man become free for his humanity and find his worth in it, the humanism differs according to one's conception of the 'freedom' and 'nature' of man. So too are there various paths towards the realization of such conceptions'.[45] It is only in this sense that the thinking of Marx and Sartre, which does not require a return to antiquity, can be defined as humanist. Now, according to Heidegger, 'however different, these forms of humanism may be in purpose and in principle, in the mode and means of their respective realizations, and in the form of their teaching, they nonetheless all agree in this, that the *humanitas* of *homo humanus* is determined with regard to an already established interpretation of nature, history, the world, the ground of the world, that is, of being as a whole'.[46] In other words, '[e]very humanism is either grounded on metaphysics or is itself made to be the ground of metaphysics', that is, it raises the question of the definition of man in a way that is not sufficiently radical, or rather without asking the question of whether this entity belongs to being. However, according to Heidegger, it is not the question of the essence of man that is misleading, but the way in which it is addressed. It is legitimate, in fact, for Heidegger to care that man is human and not inhuman, but the essence of man must be experienced in a more initial way, that is, as ek-sistence, that is, 'an ecstatic inherence in the truth of Being'.[47] Therefore, it is not so much man characterised by absolute freedom that is worthy, but rather that thought capable of thinking Being and its way of involving man, allowing him to exist. If 'the great conquest of the Renaissance', the one that imposes on us 'an obligation of eternal gratitude', was that of bringing serious

44 Ibid., p. 201.
45 Ibid., p. 202.
46 Ibid.
47 Ibid., p. 203.

and perfect knowledge of man and humanity into reality, it is to Burckhardt that we owe the conviction that 'the loftiest conception on this subject was uttered by Pico della Mirandola in his speech on the dignity of man',[48] which the Swiss historian considers 'one of the noblest bequests of the great age'.[49] It was here that, together with the Renaissance, humanism as the doctrine that considers man to be an entity of indefinite nature, worthy for his freedom and ability to define himself, was also invented. It is significant that this invention took place in a year which, as Habermann argues, was fundamental for the birth of the very category of modernity.

It is to this grand narration that Sartre and Heidegger implicitly refer when they discuss, at a distance, the possibility of giving meaning to humanism. The node of contention concerns the definition of man as free subjectivity. This different way to understand subjectivity is also closely connected with the role attributed to humanism, more generally understood as *Bildung*. Sartre does not invite us to return to the ancient world, but rather understands literature as 'the subjectivity of a society in permanent revolution'.[50] Heidegger, in a certain sense, invites thought to look back to its Greek origins, identifying in this step backwards the possibility of a new beginning that is able to conceive of Being and its multiple modes of manifestation. This is where the course of modernity is played. By inventing the Renaissance and codifying humanism, Burckhardt also determined how modernity was revived or criticised.

---

48 Ibid., p. 184.

49 Ibid.

50 Jean Paul Sartre, *Qu'est-ce que la littérature?* (Paris: Gallimard, 1948), p. 163 (my translation).

Stefano Petrucciani*

MARX'S CONCEPTION OF HUMAN NATURE

Abstract

The debate on the Marxian conception of human nature has a long history, within which the theme of human nature is obviously intertwined with that, closely related to it, of the essence of man. As is known, in the Sixth Thesis on Feuerbach, Marx writes that 'the human essence is not something abstract that is immanent to the individual. In reality, it is the whole of social relations'. From here, the question that some scholars have asked is: is there space in Marxian thought for a conception of human nature, or do the historical-social transformations to which this nature is subject make it impossible to theorize it? The thesis of the impossibility of a Marxist theory of human nature has been convincingly refuted by an acute British Marxist, Norman Geras, in a small book entitled: *Marx and Human Nature: Refutation of a Legend* (1983). The paper will therefore examine the main features of the Marxian conception of human nature. For Marx the fundamental traits that distinguish men from other animals are the ability to work and the ability to speak. An essential characteristic of man is also his sociality because, according to Marx, language and work are necessarily social activities. Since work is what (together with language) defines the humanity of men, then it follows that one of the most interesting (and least recognized) human characteristics is the need to work, that is the need to be active and to objectify one's personality.

Keywords

Marxism – Human Nature – Work – Anthropology – History

* Sapienza – Università di Roma.

1. *Humanity, Work and Language*

The discussion on the Marxian conception of human nature has a long history, within which the theme is intertwined with that, closely linked to it, of the essence of man. As is known, in the Sixth *Thesis on Feuerbach* Marx writes that 'the essence of man is no abstraction inherent in each single individual. In its reality it is the ensemble of the social relations'. Starting from here, some scholars have asked themselves the following question: is there space, in Marxian thought, for a conception of human nature, or do the historical-social transformations to which this nature is subject make its univocal theorization impossible? The thesis of the impossibility of a Marxist theory of human nature was persuasively refuted by an acute British Marxist, Norman Geras, in a small 1983 book entitled: *Marx and Human Nature. Refutation of a Legend* (Verso Books 1983). Accepting Geras' thesis, I will try in this text to identify what the main characteristics of human nature are in the Marxian conception. I will do this by taking *German Ideology* as my starting point, that is, the text where Marx achieves results that he will continue to consider valid even later. Instead, I will leave aside the reflections of the younger Marx, and in particular those found in the *Economic and Philosophical Manuscripts of 1844*, because in them there is a very complex reflection that would require an analysis entirely dedicated to it. I will try to show what are the peculiar elements of the vision that Marx developed starting from 1845 and what problems it leaves open.

Since, following Feuerbach, Marx develops his thought in antithesis to the idealistic tradition, it can be said that his conception of man is first and foremost naturalistic and materialistic. Marx always attaches great importance to the bodily, sensible and material dimension of the individual, neglected or devalued by spiritualistic philosophies. Men are primarily animals that live in a natural environment, but they distinguish themselves from other animals by a decisive characteristic: they are animals that produce their means of subsistence, that is, they are working animals. 'Men can be distinguished from animals by consciousness, by religion or anything else you like. They themselves begin to distinguish themselves from animals as soon as they begin to produce their means of subsistence, a step which is conditioned by their physical organisation. By producing their means of subsistence men are indirectly producing their material life'. In short, what makes men human according to Marx is not, as for Aristotle and the noble philosophical tradition, the possession of logos, that is, language or reason, but the activity of producing. It is certainly also true for Marx that human individuals are animals capable not only of producing, but also of speaking. From his point of view too, it can be stated that men are men by virtue of work and language. But there is no perfect equality between these two terms, because Marx emphasizes how conscience, which in men is one with language, is something that in a certain way derives from working, reproductive and relational practices. 'Language is as old as consciousness, language is practical, real consciousness that exists for other men as well, and only therefore does it also exist for me; language, like consciousness, only arises from the need, the necessity, of intercourse with other men'.[1] It therefore distinguishes man from the animal because, Marx writes, 'the animal does not "*relate*" itself to anything, it does not "*relate*" itself at all. For the animal its relation to others does not exist as a relation'.[2] 'Consciousness is, therefore, from the very beginning a social product, and remains so as long as men exist at all'.[3]

Furthermore, with work, the two fundamental characteristics that in a Marxian way can be attributed to human nature are given, namely its sociality and its historicity. If Marx does not adhere to the definition of man as *zoon logon echon*, because for him work has some priority over language, he instead fully endorses the thesis of man as *zoon politikon*:

> Man is a *zoon politikon* in the most literal sense: he is not only a social animal, but an animal that can isolate itself only within society. Production by an isolated individual outside society – something rare, which might occur when a civilised person already dynamically in

1 Karl Marx and Friedrich Engels, *The German Ideology*, in *Marx and Engels Collected Works*, Vol. 5 (London: Lawrence & Wishart, 2010), p. 44.
2 Ibid.
3 Ibid.

> possession of the social forces is accidentally cast into the wilderness – is just as preposterous as the development of language without individuals who live together and speak to one another.[4]

The theme also returns in the First Volume of *Capital* where Marx writes that '[…] man is, if not as Aristotle contends, a political, at all events a social animal'. And in a footnote, he specifies: 'Strictly, Aristotle's definition is that man is by nature a town-citizen. This is quite as characteristic of ancient classical society as Franklin's definition of man, as a tool-making animal, is characteristic of Yankeedom'.[5]

An essential characteristic of man is therefore his sociality because, according to Marx, language and work are necessarily social activities: a language without a social relationship is not possible just as an individual who works alone is not possible (as Locke or the various 'robinsonatas' represent it).

However, the historicity of human societies also derives from work. In fact, working means satisfying a need mediately, with the interposition of a tool between man and the environment (while the animal hunts its prey and eats it immediately, that is, without interposing a medium). But, if the first cycle of satisfying needs involves first of all the production of the tool (for example the chipped stone) and then its use to obtain something, in the second cycle we already find the tool ready, and therefore we can produce a bit more, and we have therefore already emerged from repetition and entered the historical novelty. But in history it is not only the environment that changes, but man himself, who not only acquires new tools, but, what is even more important to Marx, new needs: 'the satisfaction of the first need, the action of satisfying and the instrument of satisfaction which has been acquired, leads to new needs; and this creation of new needs is the first historical act'.[6] History changes the way of being of individuals and their way of

4 Karl Marx, *Economic Manuscripts of 1857–1858*, in *Marx and Engels Collected Works*, Vol. 28 (London: Lawrence & Wishart, 2010), p. 18.

5 Karl Marx, *Capital Volume 1*, in *Marx and Engels Collected Works*, Vol. 35 (London: Lawrence & Wishart, 2010), p. 331, n. 4.

6 Karl Marx and Friedrich Engels, *The German Ideology*, p. 42.

desiring, and therefore we could say that human nature is nothing more than a sort of permanent basis that is realized in always different historical configurations.

2. *Historicity of Human Nature and Critique of Negative Anthropology*

This point is well defined in a note from the First Volume of *Capital*, where Marx clearly distinguishes 'human nature in general' and 'human nature as modified in each historical epoch'.[7] A peculiarity of the Marxian approach therefore is undoubtedly that of decisively insisting on the historical modifiability of human needs, desires and passions, a peculiarity that opposes him to that long tradition of modern philosophy which, from Machiavelli and Hobbes up to Karl Löwith, insisted rather on the idea that, despite the change in historical eras, the substance of human nature (or condition) remains fundamentally the same. Marx, like Rousseau, and like most critics of society, is of the opposite opinion; he believes that the change in social conditions can also imply a change in the configuration of human passions, as emerges very clearly from the polemic that, in *German Ideology*, he conducts against Max Stirner. In fact, we read in a passage that is later deleted in the manuscript:

> Since they attack the material basis on which the hitherto inevitable fixedness of desires and ideas depended, the communists are the only people through whose historical activity the liquefaction of the fixed desires and ideas is in fact brought about and ceases to be an impotent moral injunction, as it was up to now with all moralists 'down to' Stirner. Communist organisation has a twofold effect on the desires produced in the individual by present-day relations; some of these desires – namely desires which exist under all relations, and only change their form and direction under different social relations—are merely altered by the communist social system, for they are given the opportunity to develop normally: but others – namely those originating solely in a particular society, under particular conditions of [production] and intercourse – are totally deprived of their conditions

7 Karl Marx, *Capital Volume 1*, p. 605, n. 7.

> of existence. Which [of the desires] will be merely changed and [which eliminated] in a communist [society] can [only be determined in a practical] way, by [changing the real], actual ['desires', and not by making comparisons with earlier historical conditions]. The two expressions: ['fixed' and 'desires'], which we [have just used in order to be able] to disprove [this 'unique' fact of] Stirner's, [are of course] quite inappropriate. […] A desire is already by its mere existence something 'fixed', and it can occur only to Saint Max and his like not to allow his sex instinct, for instance, to become 'fixed'; it is that already and will cease to be fixed only as the result of castration or impotence. Each need, which forms the basis of a 'desire', is likewise something 'fixed', and try as he may Saint Max cannot abolish this 'fixedness' and for example contrive to free himself from the necessity of eating within 'fixed' periods of time. The communists have no intention of abolishing the fixedness of their desires and needs, an intention which Stirner, immersed in his world of fancy, ascribes to them and to all other men; they only strive to achieve an organisation of production and intercourse which will make possible the normal satisfaction of all needs, i.e., a satisfaction which is limited only by the needs themselves.[8]

For Marx, human individuals are no more bearers of selfish passions than they are of altruistic feelings. Their dispositions in one sense or the other basically depend on the way their social relations are structured. Therefore he, always arguing against what he ironically calls Saint Max, can write about communism:

> […] the communists do not oppose egoism to selflessness or selflessness to egoism, nor do they express this contradiction theoretically either in its sentimental or in its highflown ideological form; they rather demonstrate its material source, with which it disappears of itself. The communists do not preach morality at all, as Stirner does so extensively. They do not put to people the moral demand: love one another, do not be egoists, etc.; on the contrary, they are very well aware that egoism, just as much as selflessness, is in definite circumstances a necessary form of the self-assertion of individuals.[9]

8 Karl Marx and Friedrich Engels, *The German Ideology*, pp. 255–6, footnote.

9 Ibid., p. 247.

There is therefore no space in Marx for the theorization of a drive for self-affirmation, or, to quote Kant, of an 'unsociable sociability' which would be a structural characteristic of human individuals. For Marx, the Hobbesian desire for superiority, or the Hegelian desire for asymmetrical recognition, are nothing but the fruit of an internalization of social antagonisms, destined to be overcome with them.

The Hobbesian bellum omnium contra omnes (to which Marx often refers), far from photographing the situation in which human individuals would find themselves if they were not kept in check by the norms of the State, has no other value for Marx than that of describing the condition that characterizes individuals engaged in the competitive struggle that characterizes mercantile-bourgeois society.

3. *Human Creativity and Work as a Need*

Alongside the rejection of any perspective of negative anthropology, another well-known characteristic of Marxian anthropology is the emphasis of the role that activity, work and otherwise, has in human life, that is, the objectification of one's personality in the external world. As was highlighted at the time by Erich Fromm, in his stimulating study of the Marxian conception of man,[10] human work is not only a means to achieve certain results, but is also an end in itself, because the individual is realized in the activity and objectification of his personality. For Marx, as Sean Sayers, a notable British Marxist, wrote, we as human beings are not just bearers of needs and desires, or passive consumers. We are rather 'active – productive and creative – beings. We take satisfaction in actively exercising our abilities, overcoming obstacles and being productive'.[11]

The activity that leaves a trace in the external world is a fundamental dimension of humanity because through it (whatever the objective

10 Erich Fromm, *Marx's Concept of Man* (New York: Unger, 1961).

11 Sean Sayers, 'Why Work? Marx and Human Nature', *Science & Society*, 69.4 (2005), 606–16.

results it produces) a double consequence is given rise: on the one hand the human personality becomes aware of itself and confirms itself by objectifying itself in something that is outside of it, that is, by marking a trace in the objective world. On the other hand, this objectification is also a factor of mutual recognition, that is, it makes my personality objectively visible to others. Very appropriately, Sayers notes how Marx's theses are in clear continuity with those of Hegel, and cites in this regard a fine passage from Hegelian *Aesthetics* where we read that man achieves self-awareness in two ways, both theoretically and through practical activity.

> Man brings himself before himself by *practical* activity, since he has the impulse, in whatever is directly given to him, in what is present to him externally, to produce himself and therein equally to recognize himself. This aim he achieves by altering external things whereon he impresses the seal of his inner being and in which he now finds again his own characteristics. Man does this in order, as a free subject, to strip the external world of its inflexible foreignness and to enjoy in the shape of things only an external realization of himself. Even a child's first impulse involves this practical alteration of external things; a boy throws stones into the river and now marvels at the circles drawn in the water as an effect in which he gains an intuition of something that is his own doing. This need runs through the most diversiform phenomena up to that mode of self-production in external things which is present in the work of art.[12]

The objectifications of civilization, from the most elementary to the artistic, are rooted in this original need for the active expression of the human personality. Marx follows Hegel in this way of seeing, comments Sayers, and in fact understands work not only as a means to satisfy needs, but as an end in itself. Marxianly it can therefore be stated that one of the most interesting (and often unknown) human characteristics is the need to work, understood as the need to be active and to objectify one's personality.

Marx supports exactly this thesis, when, in the *Economic Manuscripts of 1857-1858* he discusses Adam Smith's argument according to which the value of the goods produced depends on the

12 Georg W. F. Hegel, *Aesthetics* (Oxford: Clarendon Press, 1975), p. 31.

work incorporated in them because work is what they really cost, and that is the producer's renunciation of rest or carrying out the occupations he likes most. For Smith, to produce something the worker must necessarily sacrifice a certain portion 'of his ease, his liberty, and his happiness' and this is the reason why goods are normally exchanged due to the work that was necessary to produce them. However, Marx disputes the Smithian equation between work and sacrifice. The author of the *Wealth of Nations* 'is right that in its historical forms of slave labour, serf labour and wage labour, work is always repulsive and always appears as externally imposed, forced labour, and as against that not-work as "liberty and happiness"'. But Smith's considerations cannot claim to have a more general validity, because he 'has only the slaves of capital in mind. E.g. not even the skilled craftsman of the Middle Ages can be subsumed under his definition'.[13]

Although Marx's intent, as he himself declares, is not to discuss the 'philosophical' vision that Smith has of work, nevertheless he emphasizes that Smith's error, whose exact words he quotes, lies in not understanding that 'the individual "in his ordinary state of health, strength, spirits, skill, dexterity" also needs a normal portion of labour and the transcendence of "rest"'. For Marx it is not at all true that 'rest' purely and simply identifies with 'freedom' and 'happiness'. Therefore, his reasoning continues thus:

> Certainly, the volume of labour itself appears to be externally determined by the aim to be attained and the obstacles to its attainment which have to be overcome by labour. But equally A. Smith has no inkling that the overcoming of these obstacles is in itself a manifestation of freedom – and, moreover, that the external aims are [thereby] stripped of their character as merely external natural necessity, and become posited as aims which only the individual himself posits, that they are therefore posited as self-realisation, objectification of the subject, and thus real freedom, whose action is precisely work.[14]

---

13 Karl Marx, *Economic Manuscripts of 1857–1858*, in *Marx and Engels Collected Works*, Vol. 28 (London: Lawrence & Wishart, 2010), p. 530.

14 Ibid.

Marx's thesis is therefore that non-work is not at all a condition to idealize; and that on the contrary work can be gratifying and self-realizing, 'which in no way implies that work is pure fun, pure amusement, as in Fourier's childishly naive conception'.[15] But work can be self-realizing only if certain conditions are met: 'Work involved in material production can achieve this character only if (1) its social character is posited; (2) if it is of a scientific character and simultaneously general [in its application], and not the exertion of the worker as a natural force drilled in a particular way, but as a subject, which appears in the production process not in a merely natural, spontaneous form, but as an activity controlling all natural forces'.[16]

Beyond these interesting considerations, what is important to underline is that, if one starts from a complex and articulated vision of work like the one found in the mature Marx, the sharp antithesis between alienated work in capitalist society and liberated work in socialist society is no longer maintainable: in fact, as the 'socialist' work is not pure self-realization, but also submission to a necessity (as Marx explains in the Third Volume of *Capital*[17]), so, despite what Marx says, work in capitalist society can also be much more than mere alienation, because it is also, certainly to a very variable extent, objectification of the worker's personality.

4. *Problems of the Marxian Conception*

From the aspects that we have briefly reviewed so far, it can be inferred that the Marxian conception is oriented towards an active, positive and creative humanism, which precisely as such tends substantially to exorcise the dimensions of negativity and irrationality that characterize the human condition. The Marxian conception of human nature, therefore, has been in many aspects questioned by scholars and thinkers who, while sensitive to Marxian arguments,

15 Ibid.

16 Ibid.

17 Karl Marx, *Capital Volume 3* in *Marx and Engels Collected Works*, Vol. 37 (London: Lawrence & Wishart, 2010), p. 837.

have also highlighted its shortcomings. Let us dwell briefly on some of the problems that have emerged, and whose highlighting can be useful for developing a conception of man that, while preserving some aspects of the Marxian one, integrates it and makes it more comprehensive.

A first problematic point concerns, as we have already begun to glimpse, the absence in the Marxian vision of any consideration regarding the drives for self-affirmation and their conflictual implications. The line of thought that on this theme unites Hobbes and Kant, but reaches Hegel and also touches Rousseau, cannot simply be discarded. If human individuals are beings rich in needs and desiring (and the more developed they are, the more needs they have) these dimensions cannot but also place them in conflict with each other. The reasons for conflict, moreover, are not only ascribable to class society: there is conflict between the sexes, between generations, between cultures, between ethnic groups. It is therefore difficult to think that, by changing the social organization, individuals are animated only by pro-social drives and not also by selfish and antisocial drives. The thesis according to which their mutual attitude depends exclusively on the way their social relations are structured seems too radical.

Another very important aspect that has been emphasized by great readers of Marx and critics of Marxism is what in summary we could define as the underestimation of the imaginary and symbolic dimension in the definition of what are structural characteristics of the way of being of humanity. I refer here, for brevity, to two great intellectuals who have seriously engaged with this theme, Cornelius Castoriadis and Ernesto De Martino. Human activities of reproduction of life, this is the strong point that we can draw from these authors, must not only function materially. Or rather, precisely because they are human and not animal activities, they function only if they are inscribed in symbolic horizons of meaning; meaning that for human individuals is not simply given (as twentieth-century existentialism has explained to us) but must be constructed, in a way that is not predetermined, with the use of imaginary and symbolic resources.

Societies, Castoriadis[18] argues, could not solve even the most elementary problems if they did not at the same time give answer to questions of meaning, building symbolic universes that are not explainable only on the basis of functional or rational considerations. Ernesto De Martino, for his part, directly engages with the Marxian question of the essence of man, sustaining that certainly it can be conceived as 'activity', as 'going beyond', but in turn this 'going beyond' is not understood if one does not comprehend the function of symbolism as it is present in myths, religions or other dimensions of human culture. If not inscribed within a symbolic regime of protection and salvation, man would not even have been able to develop the simplest forms of work and production: if he did not have this Umweg, De Martino writes, the human individual 'would have exhausted himself, suffocated by anguish, at the first attempt to manufacture an amigdala or a digging stick'.[19] Said in other words: the primordial necessity of human individuals is not only that of reproducing their life with work, but also that of filling with symbolic constructions an open and empty horizon of meaning in which they would risk losing themselves including their role as producers. Man is not only a builder of tools but also a fabricator of symbols.

On the theme of the fragility, precarity and dependency of human nature, as a limit of a humanistic perspective all creative and positive, it has been insisted from another perspective as well. In his interesting study on materialism a great intellectual and philologist like Sebastiano Timpanaro[20] has noted how the Marxian emphasis on historical transformation runs the risk of obscuring the awareness of the permanent (though in always different modalities) dependence of man on physical nature and on his corporeal constitution, characterized by those dimensions of mortality, precarity and fragility that condition individual lives (and therefore also social dynamics) despite any historical change. Whether nourished by existentialism, as in the case of De Martino, or by 'leopardism',

18 See Cornelius Castoriadis, *The Imaginary Institution of Society* (Cambridge, MA: The Mit Press, 1987).

19 Ernesto De Martino, *La fine del mondo*, ed. by Clara Gallini (Turin: Einaudi, 1977), p. 448.

20 See Sebastiano Timpanaro, *On materialism* (London: Verso, 1985).

as in the case of Timpanaro, the twentieth-century re-readings of Marxian anthropology tend thus to attenuate its positive humanism and to bring attention instead to those dimensions of negativity that have been explored far and wide in post-Hegelian philosophy and in particular in that of the twentieth century.

Irene Viparelli*

HUMANISM AND NATURALISM IN MARX'S *1844 MANUSCRIPTS*

Notes on Kohei Saito's Ecological Reading

Abstract

In *Marx's Ecosocialism* (2018), the Japanese philosopher Kohei Saito develops a distinctive ecological reading of the *Economic and Philosophical Manuscripts of 1844*. He interprets the young Marx's idea of the unity of humanism and naturalism as an early expression of the metabolic rift thesis – the principle that, for ecosocialists, forms the ecological core of Marx's critique of political economy. This chapter critically examines Saito's interpretation of the *Manuscripts*, drawing attention to its internal contradictions. It argues that these inconsistencies stem from Saito's hermeneutic framework, revealing the difficulties inherent in approaching the young Marx's writings from an ecological standpoint.

Keywords

Marx's *1844 Manuscripts* – Humanism and Naturalism – Ecosocialism – Degrowth Communism – Kohei Saito

* Universidade de Évora.
FCT research has been financed by Research Center in Political Science UIDB/00758 University of Minho/University of Évora and supported by the Portuguese Foundation for Science and Technology (FCT) and the Portuguese Ministry of Education and Science through national funds. The present article is a translation – carried out by the author with the assistance of ChatGPT – of a text originally published in Portuguese in the volume *Política, Direito e Sociedade. Estudos em homenagem ao Professor Silvério da Rocha-Cunha*, ed. by Evanthia Balla, Irene Viparelli, Paulo Vitorino Fontes and Rafael Franco Vasques (Vila Nova Famalição: Humus, 2025, pp. 159–80).

1. *The Identity of Humanism and Naturalism: An Ecological Principle?*

The 1932 publication of the *Economic and Philosophical Manuscripts of 1844* sparked a well-known debate about Marx's humanism. By foregrounding alienated labour and humanism, the *Manuscripts* invited a reinterpretation of Marx as a humanist thinker – distinct from both theoretical economism and the harsh political practices of the Soviet regime. In the 1960s, however, Althusser's well-known thesis of 'theoretical anti-humanism' sharply challenged this interpretive framework, suggesting that humanist readings of Marx were ultimately not far from the same economism they aimed to oppose.[2] Thus, since their publication and throughout the 20th century, the *Economic and Philosophical Manuscripts of 1844* have remained at the heart of intense debate, whose political stakes centred on the relationship between Western communists and Stalinism.

What relevance do Marx's early reflections have today, in a contemporary context seemingly far removed from these historical confrontations? In 2017, the Japanese philosopher Kohei Saito offered an original ecological reading of the *Manuscripts* in the first chapter of his book *Marx's Ecosocialism*, titled 'The Alienation of Nature as the Emergence of the Modern'.[3] The principle of the identity between humanism and naturalism, as set out in the *Manuscripts*, is read by Saito as a remarkable foreshadowing of the metabolic rift thesis – a thesis Marx developed in his later works and which should represent the true ecological core of Marxian theory. In Saito's words:

> In 1844 [...] Marx sees the reason for the emergence of modern alienated life in a radical dissolution of the original unity between humans and nature. In other words, capitalism is fundamentally characterized

2 See Louis Althusser, *Umanesimo e Stalinismo* (Bari: De Donato, 1973), p. 105.

3 Kohei Saito, *Karl Marx's Ecosocialism. Capitalism, nature, and the Unfinished Critique of Political Economy* (New York: Monthly Review Press, 2017) (from now on, we will use the acronym ME, followed by the page number, directly in the body of the text).

> by alienation of nature and a distorted relationship between humans and nature. Accordingly, he envisions the emancipatory idea of 'humanism = naturalism' as a project of reestablishing the unity between humanity and nature against capitalist alienation. (ME 14)

Thus, 'in contrast to the earlier debates between "humanist" and "scientific" Marxists about the philosophical concept of "alienation"' (ME, 26), for Saito, the *Manuscripts* mark the first clear statement of the 'ecological' idea that human emancipation and environmental sustainability are fundamentally linked.

Saito's hermeneutic proposal aligns with the latest studies on ecosocialism, building on the 'ecological interpretations' of Marxian political economy by John Bellamy Foster and Paul Burkett.[4] These authors, who self-identify as 'second stage' ecosocialists, reject the representation of a Promethean and productivist Marx proposed by the 'first generation' ecosocialists.[5] They offer us a different vision of Marx – one deeply engaged with ecological issues and fully aware of the fundamental conflict between capital's drive for valorisation and the reproduction of life's material conditions. According to these authors, ecological and socialist thought doesn't emerge by distancing itself from Marx or trying to surpass the limits of his theory; rather, it does so precisely by embracing his historical-dialectical method as its epistemological foundation.

The publication of Marx's study notebooks on the natural sciences by MEGA² has undoubtedly played a key role in shaping this new image of Marx.[6] As Saito rightly points out, these notebooks prove that from the 1860s onward, Marx systematically studied the natural

4 Cf. Paul Burkett, *Marx and Nature: A Red and Green Perspective* (Chicago: Haymarket Books, 2014) and John Bellamy Foster, *Marx's Ecology: Materialism and Nature* (New York: Monthly Review Press, 2000), e-book.

5 Among the socialists of the first stage, Saito cites, in particular, Ted Benton, André Gorz, Michael Löwy, James O'Connor and Alain Lipietz.

6 There is an extensive bibliography on MEGA²'s editorial project. In particular, we recommend reading the first part of Giovanni Sgro's book *MEGA–Marx: Studi sulla edizione e sulla recezione di Marx in Germania e in Italia* (Naples–Salerno: Orthotes, 2016), dedicated to 'La Marx–Engels Gesamtausgabe (MEGA²)'.

sciences – chemistry, geology, agronomy, botany, and mineralogy – critically engaging with them. Marx paid particular attention to the 'ecological' impacts of large-scale agriculture on soil fertility. Notably, he adopted the pessimistic views of chemist Justus von Liebig regarding the 'second agricultural revolution' of his time. The attempt to counteract soil depletion caused by intensive farming through the use of fertilizers – mainly guano imported from distant regions – instead of restoring soil fertility, only worsened and expanded the metabolic disruption between humans and nature. Foster writes:

> Under the influence of Liebig, whom he studied attentively – making extensive extracts from Liebig's work in his scientific notebooks – Marx was to develop a systematic critique of capitalist 'exploitation' (in the sense of robbery, that is, failing to maintain the means of reproduction) of the soil.[7]

The concept of 'metabolism' thus underpins an ecological reinterpretation of Marx's critique of political economy.[8] Labour, as a uniquely human mode of interaction with nature, must establish a 'metabolic balance' in each historical period to ensure the reproduction of the conditions necessary both for human life and the natural environment. However, the capitalist mode of production, by turning all of nature into a commodity – a means to valorise capital – has created conditions for its unlimited exploitation, thereby disrupting the natural processes of regeneration. This has caused an

---

7 J. B. Foster, *Marx's Ecology*.

8 The concept of 'social metabolism', developed by Istvan Mészáros in his books *Beyond Capital* (New York: NYU Press, 2010) and *The Necessity of Social Control* (New York: NYU Press, 2015), is taken up by these authors as the theoretical basis for their ecosocialist reading of Marx. For Saito, the 'rupture of the metabolic balance' between man and nature is integrated by the 'spatial rupture' between the city and the countryside and the 'temporal rupture' between the time of production and the time of reproduction of nature: 'As is obvious from the slow formation of soil nutrients and fossil fuels and the accelerating circulation of capital, there emerges a rift between nature's time and capital's time'. Kohei Saito, *Marx in the Anthropocene* (Cambridge: Cambridge University Press, 2022), p. 27.

irreparable 'rupture' in the metabolic balance, triggering a process of progressive destruction of the essential conditions for life.[9]

In short, publishing the *Notebooks on the Natural Sciences* brings forth an 'ecosocialist Marx' – one who, far from any Promethean optimism, becomes increasingly aware of capitalism's inherent drive to devastate nature.

With *Marx's Ecosocialism*, Saito seeks to deepen ecological understanding of Marxian theory by offering a vital contribution: an ecological reading of Marx's entire body of work, grounded in a thorough and systematic analysis of his natural science notebooks published in MEGA². In doing so, Saito pushes ecosocialism beyond its original scope; it's no longer just about highlighting the ecological dimension in Marx's 'mature' texts, but about advancing a far more radical thesis on an ecological core of marxian critique of political economy. Saito asserts: 'I maintain that *it not possible to comprehend the full scope of his critique of political economy if one ignores its ecological dimension*'. (ME, 14)

Could the Anthropocene provide the basis for a new approach to Marx's theory – one capable of reviving scientific socialism through ecological theory, after the major political issues of the last century have passed? Within the limited scope of this paper, we do not aim to answer such a complex question. Rather, we restrict ourselves to reflecting on the interpretation Saito offers of the principle, formulated by Marx in the *Economic and Philosophical Manuscripts of 1844*, concerning the identity of humanism and naturalism. In particular, we seek to highlight certain problematic aspects of Saito's perspective that, in our view, provide a starting point for a critical

---

9 Foster demonstrates the validity of this reading by presenting the two specific passages in *Capital* in which Marx explicitly addresses the issue of metabolism. These are, in particular, chapter 47 of volume III – 'Genesis of capitalist land rent', and the analysis of 'large-scale' industrial and agricultural production, developed in volume 1. Foster comments: 'What is common to both of these passages from Marx's *Capital* [...] is the central theoretical concept of a "rift" in the "metabolic interaction between man and the earth", that is, the "social metabolism prescribed by the natural laws of life", through the robbing of the soil of it constituent elements, requiring its "systematic restoration"'. Foster, *Marx's Ecology*.

interrogation of his central thesis – namely, the idea of *degrowth communism* developed in his later works.

This paper is organised into four sections. We begin with a brief summary of Marx's principle of the identity between humanism and naturalism, as articulated in the *Manuscripts*. We then examine Saito's interpretation, drawing attention to what we consider its most problematic elements. In the final section, we put forward a conjecture about the roots of the inconsistencies in Saito's analysis, with the aim of fostering critical engagement with the 'degrowth communism' hypothesis.

2. *On the Identity of Humanism and Naturalism*

Before delving into Saito's interpretation of the ecological core of the *Economic-Philosophical Manuscripts of 1844*, it is worth beginning with a brief introduction to the *Manuscripts* themselves, and in particular to the principle of the identity between humanism and naturalism. Written in the first half of 1844, the *Manuscripts* are composed of three notebooks. In the first, Marx offers a critical analysis of the three central categories of classical political economy – wages, profit and land rent – in an effort to demonstrate their inner contradictions on the basis of political economy's own premises. In the second part of this first notebook, Marx adopts Feuerbach's method of religious critique to formulate his well-known theory of 'alienated labour': just as in religion, where humanity's essential capacities are projected outward to form a transcendent being – God – so too, in alienated labour, human faculties are estranged from the worker and objectified in the form of private property.

In the young Marx's view, labour constitutes humanity's universal essence – our fundamentally human way of relating to nature. Through labour, humans transform nature into their 'inorganic body', shaping it according to social needs and creating an 'objective world' tailored to those needs. This conception of man as an 'essentially productive being' carries a twofold implication. First, in opposition to Hegel and the idealists, it affirms man's natural, tangible, and sensory existence, which depends on an intrinsic connection with nature.

Second, and this time in contrast to Feuerbach, it portrays man not merely as a 'sentient being' but foremost as a sentient 'activity' – a praxis that humanises nature by adapting it to human purposes. Marx asserts: 'It is therefore in his fashioning of the objective that man really proves himself to be a species-being. Such production is his active species-life. Through it nature appears as his work and his reality'.[10] In other words, man does not merely produce to meet the conditions for individual and species survival, but also – above and beyond the needs of reproduction – in a free and universal manner, aligned with human social needs and 'in accordance with the laws of beauty'. (EPM44, 329)

Because labour is humanity's fundamental activity, the young Marx understands its alienation as a profound loss of our very essence – a process that dehumanises by turning the creation of an objective world into estrangement from all of nature. In alienated labour, both the product and the activity itself are taken from the worker, becoming the property of someone else. This alienation extends to relationships with others and with one's own species, producing a complete inversion of means and ends, of the universal and the particular. Human beings' species-activity – labour – no longer expresses the objectification of their universal essence, but is reduced to a mere means of satisfying their individual, immediate, and animal needs. This is why, Marx concludes, alienated labour 'transforms his advantage over animals into the disadvantage that his inorganic body, nature, is taken from him. In the same way as estranged labour reduces spontaneous and free activity to a means, it makes man's species-life a means of his physical existence'. (EPM44, 329)

The analysis of alienated labour requires a fundamental rethinking of political economy's core assumptions: the categories of private property and wage labour – typically regarded as the foundations of economic theory – are instead expressions of labour's alienation.

---

10 Karl Marx, 'Economic and Philosophical Manuscripts (1844)', in *Early Writings*, trans. Rodney Livingstone and Gregor Benton (London: Penguin Classics, 1992), p. 329 (from now on, we will use the acronym EPM44 followed by the page number, directly in the body of the text).

Marx identifies land rent and capital as the two crucial phases of human alienation. Under the regime of land rent, alienation is still somewhat masked by the personal ties between lord and land. But with the rise of capital – the climax of the alienation process – the full severance of humanity from nature is achieved, revealing private property in its most ruthless form: 'In this way the medieval saying *nulle terre sans seigneur* gives way to the modern saying *l'argent n'a pas de maitre*, which is an expression of the complete domination of dead matter over men'. (EPM44, 319)

In brief, private property's historical evolution reveals its contingent and non-essential nature, clearing the path for its abolition. Communism, then, stands as the culmination of this historical process – a movement to overcome alienation, reclaim humanity, and establish a truly human relationship with nature, one that is universal and free. Thus, it is in the third notebook of the *Manuscripts* that Marx finally lays out the principle of the identity between humanism and naturalism. As he puts it:

> This communism, as fully developed naturalism, equals humanism, and as fully developed humanism equals naturalism; it is the *genuine* resolution of the conflict between man and nature, and between man and man, the true resolution of the conflict between existence and being, between objectification and self-affirmation, between freedom and necessity, between individual and species. [...] *Society* is therefore the perfected unity in essence of man with nature, the true resurrection of nature, the realized naturalism of man and the realized humanism of nature. (EPM44, 348–9)

3. *Saito's Reading of the Principle of Identity Between Humanism and Naturalism*

Saito's fresh take on the *Economic-Philosophical Manuscripts of 1844* rests on a novel reading of the first notebook. As mentioned earlier, Marx initially critiques economic categories from within the framework of political economy itself. But then, in a sharp turn, he shifts to a critique rooted in the concept of alienated labour. This shift has prompted many interpreters to split the notebook in two, separating political economy from philosophical critique. As a result, stepping away from the political economy perspective has often been seen as a prerequisite for the philosophical critique of alienated labour.[11]

Saito insists that this supposed split isn't just wrong – it ends up making Marx's analysis look circular and self-contradictory. If Marx shows private property arises from alienated labour, then how can he explain the origin of alienated labour itself? Why does man alienate his own essence and externalise it? For Saito, this is a false problem, easily dismissed once we see that philosophical and economic critiques are deeply intertwined, each shedding light on and shaping the other. In this fresh interpretive framework, Marx's idea of alienated labour isn't just a leftover from Feuerbach's materialism – it's born directly from Marx's economic thinking. As Saito puts it: 'In contrast to the dominant tendency, I argue that the 'emergence of a theory' in Marx's notebook must be understood in a close relation to his analysis of political economy because his original theory of alienation is formulated in the process of a critique of it'. (ME, 33) As a result, Saito argues that the sections of the *Manuscripts* on land rent and capital form the core of Marx's theory of alienation. Paraphrasing Marx, Saito highlights the crucial differences between these two forms of private property: while the medieval serf was bound to the land – both being the property of the lord – capital represents the full separation of man from nature. In the pre-capitalist era, alienation appears as an absolute unity between man and nature, rendering the serf a natural, inert element – an extension of the earth, used in constructing the lord's objective world. As capital gradually takes over from land rent, land loses its inalienable status and becomes just another commodity ruled by market forces; at the same time, serf labour gives way to wage labour. The bonds that once tied serf to land and land to lord unravel, turning both worker and nature into mere exchange values – nothing more than instruments for capital accumulation, wholly indifferent to human needs.

11 Marcuse's 1932 article 'New sources for the foundation of historical materialism' is, according to Saito, the emblematic example of this perspective.

In essence, Saito views alienated labour not merely as a philosophical concept but as a historical process that embodies the full alienation of man from nature under capitalism. This reveals that even early on, Marx saw the separation of man from the material basis of life and production as central to capitalism – a clear precursor to the metabolic rift thesis. As Saito puts it: 'Only if one comprehends the estrangement in capitalist society as a dissolution of humans' original unity with the earth does it becomes evident that Marx's communist project consistently aims at a conscious rehabilitation of the unity between humans and nature'. (ME, 42)

Communism, as the realized unity of naturalism and humanism, is far more than simply the 'realisation of the human essence', as Althusser suggested. From the *Manuscripts* onward, it embodies the historically specific demand to restore the metabolic balance between man and nature, disrupted by capitalism. In this sense, communism represents the effort to establish a new, conscious, and rational relationship between humanity and the material conditions necessary for life's reproduction. Saito states:

> Marx's vision of the future society is without doubt fully consistent with his critique of modern alienated labor. It is in this *economic* sense that Marx in 1844 insists that establishment of the absolute unity of humanity and nature is the central task of communism. [...] Marx's critique of alienation of 1844 regards the 'rational' reorganization of the relationship between humans and nature as essential, and thus he envisions the idea of communism as the accomplished 'humanism = naturalism'. This is a beginning, even if it is only a beginning, of Marx's economic and ecological critique of capitalism. (ME, 43–4)

4. *The Young Marx: Between Philosophical Abstractions and Historicity – The Shortcomings of Saito's Reading*

Saito sees the discovery of alienated labour as revealing the distinctiveness of the young Marx's outlook compared to the young Hegelians, opening up a fundamentally different and radically materialist philosophical horizon. Indeed, Saito's reading of the first notebook of the *Manuscripts* –linking the concept of alienated labour with the critique of political economy – highlights a deep continuity within Marx's thought, rooted in the capacity of political economy critique to shape philosophical discourse itself. Put simply, it is the historical grounding of alienated labour, emphasized by Marx through his analysis of the historical forms of private property, that secures the essential coherence of the Marxian project. Thus, for Saito, in the *Grundrisse* Marx reaffirms the idea that alienation results from the absolute separation of the conditions of production, a concept that remains central. As he notes, 'even though Marx does not use the term 'alienation', the theoretical continuity since 1844 is quite obvious'. (ME, 47)

Saito seems to definitively move beyond the old debate about the humanism of the young Marx. On one hand, by establishing theoretical continuity, the Althusserian hypothesis of an 'epistemological break' immediately loses its foundation; on the other, humanist interpretations that limit this continuity to purely philosophical reflection are also questioned. However, later in the chapter, Saito takes an unexpected turn. Revisiting the relationship between economic critique and philosophical reflection, he adopts a stance close to Althusser's, arguing for a sharp theoretical break between the *Manuscripts* and Marx's later works – specifically the *Theses on Feuerbach* and *The German Ideology* – and reading the *Manuscripts* as grounded in a Feuerbachian and essentialist framework.

As is well known, in these texts Marx delivers a pointed critique of Feuerbach's materialism, faulting it for its narrow conception of materiality as mere sensory intuition and for failing to grasp it as the product of human praxis and transformative activity. Drawing on Marx's own arguments against Feuerbach, Saito highlights what he sees as the theoretical limitations of the *Manuscripts*, which remain, in his view, confined within a fundamentally Feuerbachian philosophical framework. Saito writes:

> Since Marx's critique was directed only against the abstract nature of idealist philosophy from his own standpoint of sensuous perception, he, together with Feuerbach, recommended overcoming these philosophical antitheses with concrete sensuous praxis and, more precisely, 'labor' that can actualize the free and universal subjectivity of human beings in the concrete objective world. Thus what Marx problematized in the

> notebooks of 1844 is essentially dependent upon the return to concrete 'sensuous perception' in labor being the *true* principle of radical materialism, and in this vein Marx demanded that human beings should first correctly *recognize* their own species-being *and then* get engaged in revolutionary praxis against alienated reality under capitalism. (EM, 54)

It is no longer the productive interplay between economic and philosophical thought that illuminates the young Marx's *Manuscripts*. Instead, the two domains are once again pulled apart, with philosophy now portrayed as the main theoretical obstacle to a historical analysis of alienation. In the final part of the chapter, Saito thus puts forward a second reading of the principle of the identity of 'humanism and naturalism' – one that, unlike the first, sees its transformative potential diminished by its entrenchment in Feuerbach's philosophy. In Saito's words:

> Even though Marx was aware that humans always relate to nature through the mediation of labor and that modern alienation deforms this relationship, his entire project of communism in 1844 was dependent upon a philosophically conceptualized idea of 'humanism = naturalism'. Since his critique of alienation still roughly identified 'capitalism' with 'the system of private property', Marx inevitably fell into a deterministic understanding of history, one that failed to carefully analyze the historical specificity of the capitalist mode of production. (EM, 57)

Saito now suggests that every element of Marx's early analysis is marked by a general 'contamination' from a 'philosophically conceptualised idea'. In this light, even the economic critique developed in the first notebook no longer reflects the deep continuity of the Marxian project, but rather reveals the young Marx's failure to grasp the specific nature of the capitalist mode of production. Alienated labour, in turn, is no longer seen as an early expression of the metabolic rift thesis, but as the product of a philosophical construct – the idea of the identity between humanism and naturalism – which leads to a deterministic, teleological, and ultimately idealist view of history.

Having thus stripped the *Manuscripts* of their anticipatory force, Saito ultimately locates the real theoretical foundation for ecological thinking in *The German Ideology*, where 'in contrast to Feuerbach, he comprehends the antagonistic relationship between humans and nature as a specific modern product that resulted from capitalist industrialization'. (EM, 59) Thus, it is no longer in the *Manuscripts*, but only in this later text that Marx would provide a firm theoretical basis for ecological thinking. Only in *The German Ideology*, in other words, would Marx decisively break from the Young Hegelians, framing the historical transformation of modes of production as the key factor in shaping the relationship between nature and social production. History – previously sidelined by the dominance of philosophical concerns – would finally move to the centre of analysis.

Saito's shift in perspective reaches a paradoxical point here: the historicity of alienation – and the specific character of capitalist alienation as distinct from its pre-capitalist forms – initially presented as key insights of the *Manuscripts*' first notebook, are now reinterpreted as the outcome of what Althusser would describe as a 'change of terrain': the product of a new, historically grounded materialism formulated in *The German Ideology*. At the same time, the *Manuscripts* are reabsorbed into the intellectual orbit of Young Hegelianism.

5. *Promethean versus Ecological Marx: Final Remarks*

In the first part of his chapter on the *Manuscripts*, Saito presents the principle of the identity between humanism and materialism as the embryonic form of the metabolic rift concept. As shown, this interpretation hinges on understanding alienated labour not merely as a philosophical notion but as a historical one – a concept shaped not only by Feuerbach's philosophy but also by the study of political economy. In the second part of the chapter, however, Saito puts forward a counterpoint, arguing that economic and historical analysis must be disentangled from the philosophical and Feuerbachian framework. This leads us to ask: how are we to understand Saito's ambivalent position regarding the young Marx's Feuerbachianism?

In our view, the answer lies in the section of the third notebook of the *Manuscripts* dedicated to the critique of Hegelian dialectics. It is there, in fact, that the young Marx articulates the connection between the historicity of political economy and the philosophical concept of alienation. As Marx writes:

> The importance of Hegel's *Phenomenology* and its final result – the dialectic of negativity as the moving and producing principle – lies in the fact that Hegel conceives the self-creation of man as a process, objectification as loss of object [*Entgegenstdndlichung*], as alienation and as supersession of this alienation; that he therefore grasps the nature of *labour* and conceives objective man – true, because real man – as the result of his *own labour*. (EPM44, 385–6)

Hegel uncovers the 'essence of labour' and, with it, the historicity of human, though only within the speculative framework of his dialectic. In this context, work is understood solely as alienation – essentially, as de-objectification. Hence, for Marx,

> Hegel adopts the standpoint of modern political economy. He sees *labour* as the *essence*, the self-confirming essence, of man; he sees only the positive and not the negative side of labour. Labour is *man's coming to be for himself* within *alienation* or as an *alienated man*. (EPM44, 386)

The materialist principle of the identity between humanism and naturalism moves beyond the limits of both Hegel and political economy by preserving the central role of alienated labour while recognising the historical and material nature of man. As Marx writes: 'Here we see how consistent naturalism or humanism differs both from idealism and materialism and is at the same time their unifying truth. We also see that only naturalism is capable of comprehending the process of world history'. (EPM44, 389) Thus, Marx's perspective in *Manuscript* diverges sharply from both Hegel and Feuerbachian philosophy. In Lukács words:

> In the *Economic and Philosophical Manuscripts*, albeit still in merely embryonic form, Marx already sets in opposition to the Hegelian mystification – and to the absurdities resulting from it – the conception of history grounded in historical and dialectical materialism. In doing

> so, he simultaneously and definitively overcomes the limitations of Feuerbach as well.[12]

In conclusion, the young Marx's critique of Hegelian dialectics – by linking alienated labour to human historicity – seems to support Saito's reading of the first notebook, where economic and philosophical critique are shown to be deeply intertwined. Defining man through labour immediately shifts the theoretical ground beyond Feuerbach, highlighting the fundamentally historical core of Marx's project. This also challenges Althusser's (and to some extent Saito's) claim about the essentialism of the *Manuscripts*. It reveals the *Manuscripts* as a crucial moment in the development of Marx's materialist and dialectical thought. So, why then does Saito overlook the critique of Hegel presented in the third notebook? Our hypothesis is that the dialectical structure of Marx's analysis reframes the identity of humanism and naturalism and undermine his own ecological interpretation.

The dialectical image of world history as the process of 'man's self-production by man' presents human history as a process in which nature is progressively subordinated to human needs. Communism, understood as the full humanisation of nature and the naturalisation of man, emerges from this historical movement. Even amid alienation, it lays the material foundation for transforming humanity into a diverse multitude of free, prosperous, and well-rounded individuals. Within this framework, the development of human productive forces is the essential condition – the *sine qua non* – for the humanisation of nature. As Marx puts it:

> The product of labour is alien to me and confronts me as an alien power, to whom does it then belong? To a being *other* than me. Who is this being? The *gods!* It is true that in early times most production – e.g. temple building, etc., in Egypt, India and Mexico – was in the service of the gods, just as the product belonged to the gods. But the gods alone were never the masters of labour. The same is true of *nature*.

12 Georg Lukács, *O jovem Marx e outros escritos de filosofia*, trans. by C. Nelson Coutinho e J. P. Netto (Rio de Janeiro: Editora UFRJ, 2009), p. 190 (my translation).

> And what a paradox it would be if the more man subjugates nature through his labour and the more divine miracles are made superfluous by the miracles of industry, the more he is forced to forgo the joy of production and the enjoyment of the product out of deference to these powers. (EPM44, 330)

The identity of humanism and materialism no longer comes across as an early formulation of the metabolic rift thesis. Rather, it points to a Promethean, productivist outlook that sits fundamentally at odds with ecological thinking. Thus, for Saito, communism is a movement aimed at restoring the disrupted metabolic balance – a 'return' to the natural limits shattered by the commodification of the world. It rejects the idea of dialectics as a linear progression and marks a radical break from the unsustainable logic behind capitalistic nature's 'theft'. At the same time, in seeking to re-establish a virtuous metabolism, this vision of communism looks back to pre-capitalist examples. Hence, in *Marx's Ecosocialism*, it is the 14th-century English yeoman – who freed himself from medieval restrictions to work his own small family farm – who embodies the anticipation of humanity in communist society. As Saito explains:

> The development of 'the free individuality of the worker' is an expression that Marx usually uses in the context of a future society established among the associated producers, but as an exception he uses it to characterize precapitalist small-scale family agriculture, where the worker can behave as 'the free proprietor of the conditions of his labor', even if it is still a limited premodern form. (EM, 49)

To sum up, the ecological reading of the principle of the identity between humanism and naturalism can only be sustained through a double – and somewhat contradictory – move. First, to present this principle as a forerunner to the metabolic rift thesis, alienation must be understood in its historical core, moving beyond the Feuerbachian philosophical framework. This is precisely what Saito sets out to do in the first part of the chapter. However, the philosophical-historical principle must be disentangled from its Hegelian foundations. As a result, Saito is – somewhat paradoxically– led to reassert the separation between the two in the chapter's second part. In doing so, he edges closer to Althusser's thesis of an epistemological break, calling for a radical 'change of ground' and a 'departure from philosophy' as necessary steps towards the foundation of scientific materialism.

In our view, Saito is not so much opposing Feuerbach or essentialism here, but rather the dialectical, Hegelian core of Marx's perspective. The inconsistencies and ambiguities in his reading of the *Manuscripts* arise from his attempt to present the principle of the identity of humanism and naturalism – as formulated by the young Marx – as the embryonic form of the ecological principle of metabolic rift. Yet in reality, this principle reflects a dialectical – and, in Saito's own view, Promethean – outlook, one that sees capitalist development as a necessary condition for communism. Our hypothesis appears to be confirmed by the evolution of Saito's thinking. In his later book, *Marx in the Anthropocene*, he ultimately abandons the aim of offering an integrated ecological reading of Marx's entire work, and instead puts forward a radically different thesis:

> Only by paying attention to Marx's theoretical crisis does it become clear why he had to simultaneously study the natural sciences *and* pre-capitalist societies while attempting to complete the subsequent volume of *Capital*. By intensively studying these theoretical fields, Marx ultimately went through another paradigm shift after 1868. It is from this perspective that Marx's letter to Vera Zasulich sent in 1881 needs to be reinterpreted as the crystallization of his non-productivist and non-Eurocentric view of the future society, which should be characterized as degrowth communism.[13]

Saito shifts the moment of Marx's 'epistemological break' away from 1845, as proposed by Althusser, and places it instead in 1868. It is only from this point onwards, he argues, that Marx – through his engagement with the natural sciences and his study of pre-capitalist and non-Western social formations – breaks decisively with the Promethean outlook of his earlier work and begins to develop a genuinely ecological perspective. In other words, it was not until

13 Saito, *Marx in the Anthropocene*, p. 8.

1868 that Marx, having distanced himself from the Eurocentrism and productivism of his youth, was able to disentangle communism from dialectical linearity and articulate the idea of a 'degrowth communism'. As Saito puts it:

> By the 1880s Marx recognized that the persistent stability of communes without economic growth is the underlying foundation for realizing sustainable and egalitarian metabolic interaction between humans and nature. This marks a clear contrast to Marx's previous negative comments on the stationary state and invariability of the Asian communes in the 1850s and even in volume I of *Capital*. This is how these two seemingly irrelevant research fields of natural science and communes prove tightly interwoven in Marx's abandonment of his earlier historical materialism.[14]

In short, Saito himself shifts his position, now presenting the principle of the identity of humanism and naturalism in Marx's early writings as a 'monist' framework that reduces nature to man's 'inorganic body' – a view fundamentally at odds with sustainability. It is only when Marx comes to terms with the fundamental dualism between humanity and nature – and recognises the need to respect nature's metabolic processes in the face of unchecked human productivity – that a truly ecological perspective begins to emerge in his thought. The idea of an 'epistemological break' in 1868 allows us to move beyond the puzzles surrounding interpretations of the young Marx, rejecting the claim that the principle of the identity between humanism and naturalism contains the seed of ecological thought. Yet, in our view, it also raises an entirely new set of challenges – particularly concerning the 'degrowth communism' hypothesis.[15] We therefore conclude not with definitive answers, but with an open question: Is the principle of humanism and naturalism – as articulated

---

14 Saito, *Marx in the Anthropocene*, p. 208.

15 The thesis of 'degrowth communism' distances Saito from Foster's perspective, who, on the contrary, is critical of degrowth theory. See the chapter 'Capitalism and Degrowth: An Impossibility Theorem' in the book *Capitalism in the Anthropocene. Ecological Ruin or Ecological Revolution* (New York: Monthly Review Press, 2022), e-book.

in the *Manuscripts* within a materialist and dialectical framework – inevitably an expression of monist Prometheanism?

Historian and geographer Jason Moore sees Foster's idea of the 'metabolic rift' as somewhat ambiguous. On one hand, it opens up new directions for developing a Marxist ecological theory; on the other, it suggests a dualistic take on the relationship between humans and nature – what he calls a 'green arithmetic' – that's basically non-dialectical and clashes with core Marxist principles. Moore aims to clear up these tensions by placing the metabolic rift within an ontological-relational and deeply dialectical reading of Marx. From this standpoint, both the dualist view of nature as separate from humanity and the monist depiction of humans as nature's creators and rulers are left behind:

> If metabolism is not an exchange between quasi-independent objects (Nature/Society) but rather a process of life-making within the web of life new possibilities emerge. A conception of a singular, internally diverse, historically variant and geographically uneven, metabolism of humanity-in-nature might allow us to chart a course beyond dualism.[16]

Within this interpretive framework, the principle of the identity between naturalism and humanism, as laid out in the *Manuscripts*, does not amount to an alleged 'Prometheanism' in the thought of the young Marx. Rather, it constitutes the very foundation of a dialectical and materialist ontology. As the Portuguese Marxist philosopher José Barata-Moura notes:

> From the standpoint of a dialectical materialist ontology developed within a Marxian (and Marxist) horizon, the 'materiality' and 'thingness' of being are not identical. Not because the domain of the 'human' is an excrescence or an addition to being (introducing unavoidable dualistic perspectives); nor because being originally dissolves into the 'human' (opening the way to various monisms of inescapable idealist character) – but because, in its own materiality, the productive sphere of the

16 Jason W. Moore, 'Metabolic rift or metabolic shift? dialectics, nature, and the world-historical method'. *Theory and Society,* 46 (2017), 285–318, https://doi.org/10.1007/s11186-017-9290-6. See also Jason W. Moore, *Capitalism in the web of life* (London and New-York: Verso, 2018).

> 'human', the human community in its multifaceted dynamics, is an intrinsic expression and mediation (worked) of being.[17]

In brief, taking Moore's reading as a starting point, Saito's thesis – which places the crucial epistemological break in 1868 as foundational for the idea of a 'communism of degrowth' – can be understood as a further development and deepening of the dualist strain already evident in Foster's concept of the metabolic rift. Would it not be, then, not the philosophical core of the young Marx's principle of identity between humanism and naturalism, but rather Saito's radicalisation of the dualistic tendency within the concept of metabolic rift, that marks the true theoretical limit to be overcome – the site of a necessary epistemological break?

17 J. Barata-Moura, *Materialismo e subjetividade. Estudos em torno de Marx* (Lisboa: 'Avante!', 1997), p. 174.

Blaise Bachofen[*]

ROUSSEAUIST PERFECTIBILITY AND DARWINIAN EVOLUTION
Crossed Issues and Methodologies[1]

Abstract

This article explores the philosophical and anthropological convergences and contrasts between Rousseau's theory of perfectibility and Darwin's theory of evolution, with particular attention to *The Descent of Man*. While Rousseau famously defends a discontinuist view of human nature as qualitatively distinct from that of animals, Darwin adopts a continuist and materialist framework. Yet a closer reading reveals an overlooked dimension of Darwin's thinking: the idea that natural selection in humans may be mediated by collective aesthetic and moral judgments, especially through mechanisms of social recognition. This perspective points to a potential tension in Darwin's account of the emergence of morality and sociability, insofar as it introduces mediations that are not easily reducible to purely mechanistic explanations. Rousseau's work, for its part, reveals a marginal yet notable interest in forms of continuity between humans and animals. Moreover, his analysis of the moral transformation of the passions and of self-love outlines a genealogical model of morality that, in some respects, resonates with Darwin's own reflections. By comparing these two conceptions of human becoming, the article sheds light on the possibility of a dialogue between philosophy and the natural sciences.

* Cergy Paris Université.

1 I would like to thank Raffaele Carbone, who organized the conference at which I had the opportunity to deliver this paper and who suggested that I work on this comparison between Rousseau and Darwin.

Keywords

Jean-Jacques Rousseau – Charles Darwin – Perfectibility – Theory of evolution – Genealogy of moral norms

Attempting to draw a connection between the theses, questions, and methods of Rousseau (1712–1778) and those of Darwin (1809–1882) is both a stimulating and perilous task. Why – and on what grounds – should one engage in a dialogue between two thinkers who lived in different centuries, who worked within distinct conceptual frameworks, and of whom the latter never once mentions the former? At first glance, there are formal analogies between their reflections on human nature. Both situate civilized man within a historical process and attempt to understand his development: Rousseau through the notion of *perfectibility* in the *Discourse on Inequality* (1755), Darwin through that of *evolution* in *The Origin of Species* (1859) and *The Descent of Man* (1871). However, there are significant differences. Rousseau's theory of perfectibility belongs to a philosophy of mind aimed at establishing the qualitative specificity of the human being, setting him apart from the animal. Darwin, by contrast, views humans as animals among others, inscribing their history within a continuous evolutionary process. Whereas Rousseau presents his hypotheses as the fruit of 'meditative' reflection,[2] Darwin carefully distances himself from philosophical speculation on human nature.

Nevertheless, some of their respective theses tend to blur these oppositions. Rousseau's work, though discontinuist in its orientation, reveals an interest, albeit a marginal one, in forms of continuity between man and animal. Darwin, for his part, formulates hypotheses on the formation of moral judgment and the genesis of sociability that suggest a speculative, even philosophical, dimension

2 See for example Jean-Jacques Rousseau, *Discourse on the Origin and the Foundations of Inequality Among Men*, or *Second Discourse* (hereafter abbreviated as *SD*), ed. and trans. by Victor Gourevitch, in *The Discourses and Other Early Political Writings* (Cambridge, UK: Cambridge University Press, 1997), Epistle Dedicatory, p. 114, Preface, pp. 125 and 127, Part I, p. 143; and *passim*.

often overlooked – one not unrelated to the reflections of major thinkers such as Rousseau on the moral transformation of the passions. A close reading of *The Descent of Man* shows that Darwin, when addressing human evolution, includes within the selection process a mediating role played by collective judgments – particularly aesthetic ones. This mode of explanation, rarely acknowledged or discussed, complicates the materialism and continuism often dogmatically associated with his name. Comparing these two conceptions of human becoming thus helps illuminate the conditions for a dialogue between philosophy and the natural sciences.

1. *A formal homology of approaches*

Rousseau and Darwin both offer a reflection on the transformation of the human being without any predetermined finality. It is primarily in this absence of teleology that a formal homology between *perfectibility* and *evolution* can be found.

Rousseau develops what he calls a 'Theory of Man'.[3] This anthropology is based on a 'natural, moral and political history' of humankind.[4] Even if, in the eighteenth century, the word 'history' could simply mean 'inquiry', Rousseau indeed integrates the idea of transformations over time, both natural (physiological) and moral or political (cultural, social, institutional). The notion of 'perfectibility' (a term that appears only once in Rousseau's corpus)[5] is outlined in the *Discourse on the Sciences and the Arts*, developed in the *Discourse on Inequality*, and provides a major source of inspiration for *Émile*. In the *First Discourse*, Rousseau writes: 'Our souls have become corrupted in proportion as our Sciences and our Arts have

3 Jean-Jacques Rousseau, *Letter to Beaumont* [1763], trans. by Christopher Kelly and Eve Grace, in *The Collected Writings of Rousseau*, 13 vols (Hanover and London: University Press of New England, 1990–2010), vol. 9, 2001, p. 31.

4 *SD*, Note X, p. 211.

5 On the historical development of the notion of 'perfectibility', see Bertrand Binoche, *Nommer l'histoire* (Paris: Éditions de l'EHESS, 2018).

advanced toward perfection'.[6] In the *Second Discourse*, he defines perfectibility as 'a faculty which, with the aid of circumstances, successively develops all the others, and resides in the species as well as in the individual, whereas an animal is at the end of a few months what it will be for the rest of its life, and its species is at the end of a thousand years what it was in the first year of those thousand'.[7] Unlike Condorcet, who speaks of an 'indefinite perfectibility',[8] involving the idea of linear and inevitable progress, Rousseau insists on ambivalence of perfectibility. He refers not to the improvement of man himself, but of his faculties – an intensification of the passions as well as a development of reason – which may lead to both corruption and progress: 'It would be sad for us to be forced to agree that this distinctive and almost unlimited faculty is the source of all of man's miseries; [...] that it is the faculty which, over the centuries, causing his enlightenment and his errors, his vices and his virtues to bloom, eventually makes him his own and Nature's tyrant'.[9]

Darwin, for his part, also presents a non-teleological conception of human becoming. Against Lamarck, he rejects the idea that this evolution obeys a final intention.[10] And, which is less well known, he contests the notion of a universal progress of human societies:

6 Rousseau, *Discourse on the Sciences and Arts*, or *First Discourse* [1750], Part I, in *The Discourses and Other Early Political Writings*, p. 9.

7 *SD*, Note X, p. 211.

8 Jean-Antoine-Nicolas de Condorcet, *Esquisse d'un tableau historique des progrès de l'esprit humain* (Paris: Masson et fils, 1822), Preface and Ninth Epoch, pp. 3 and 214.

9 *SD*, Part I, p. 141.

10 'In Lamarck's theory, transformation occurs as a result of external circumstances generating "needs": [...] the giraffe's neck lengthens so that it can feed on the high leaves of trees. For Darwin, 'spontaneous and accidental' hereditary variations within a species increase its diversity, and it is on that basis that natural selection operates: in the case of the giraffe, there are variations in neck length in a giraffe ancestor, and the environment (the height of the branches) serves as the selection filter; individuals with longer necks will reproduce more efficiently because they are better nourished [...]. In itself, this process implies no tendency toward increasing complexity, let alone perfection'. Jean-Claude Bregliano, 'Lamarck et Darwin: deux visions divergentes du monde vivant' (*Encyclopédie de l'Environnement*, 25 February 2021), online

> Many savages are in the same condition as when first discovered several centuries ago. As Mr. Bagehot has remarked, we are apt to look at progress as normal in human society; but history refutes this. The ancients did not even entertain the idea, nor do the Oriental nations at the present day. According to another high authority, Sir Henry Maine,[11] 'the greatest part of mankind has never shewn a particle of desire that its civil institutions should be improved'. Progress seems to depend on many concurrent favourable conditions, far too complex to be followed out.[12]

Yet a profound divergence must be underscored: for Rousseau, perfectibility applies to the already modern human being,[13] and he mainly defends a discontinuist thesis, whereas Darwin adopts a continuist perspective.

Just before introducing the notion of perfectibility in the *Discourse on Inequality*, Rousseau sets out a metaphysical position, and what he then says about perfectibility serves to support that position. Man is distinguished from animals not by understanding, but by freedom. 'Every animal has ideas, since it has senses, up to a point; it even combines its ideas, and in this respect man differs from the Beast only in degree'. But he then adds: 'It is [...] not so much the understanding that constitutes the specific difference between man and the other animals, as it is his property of being a free agent. Nature commands every animal, and the Beast obeys. Man experiences the same impression, but he perceives himself free to acquiesce or to resist; and it is mainly in the consciousness of this freedom that the spirituality of his soul exhibits itself'. It is immediately after this

---

ISSN 2555-0950, URL: https://www.encyclopedie-environnement.org/vivant/lamarck-darwin-deux-visions-divergentes-monde-vivant/. [Translation mine.]

11 Henry Maine, *Ancient Law* (London: John Murray, 1861), p. 22. For Mr. Bagehot's remarks, see 'The Age of Conflict', *Fortnightly Review*, 12.1 (1868), p. 452. [Darwin's note]

12 Charles Darwin, *The Descent of Man, and Selection in Relation to Sex* (hereafter abbreviated as *DM*), [2nd ed., 1874 (1st ed. 1871)], Part One, chap. V, in *The Works of Charles Darwin*, 29 vols (London: William Pickering, 1987-90), vol. 21, ed. by Paul H. Barrett and Richard B. Freeman, pp. 137–8.

13 'Modern human being', in the sense given to this term by paleoanthropology.

affirmation that he evokes perfectibility: 'But even if the difficulties surrounding all these questions left some room for disagreement about this difference between man and animal, there is another very specific property that distinguishes them, and about which there can be no argument, namely the faculty of perfecting oneself'.[14]

This faculty, empirically observable in its effects (the invention of tools, of concepts), presupposes, in his view, free will and therefore an immaterial soul. From this follows a dualist hypothesis – cautiously stated but maintained in several texts, notably in *Émile* and the *Dialogues* – and a discontinuist thesis reiterated in the *Discourse on Inequality*, notably concerning language.[15]

2. *Science versus philosophy?*

Although Rousseau and Darwin both reflect on the history of humankind and the process of hominization, their approaches and methods are therefore radically different, at least at first glance.

The question of the origin of the human being is one of the first questions Rousseau poses,[16] whereas it is among the last Darwin addresses. Darwin is cautious, even distrustful, toward philosophical speculation: he conceives of himself first and foremost as a scientist. He approaches the application of his theory of the origin of species to the human being only very late, and with caution. The thesis according to which humans belong to the same family as the great

14 *SD*, Part I, pp. 140–1.

15 'As for myself, frightened by the increasing difficulties, and *convinced of the almost demonstrated impossibility that Languages could have arisen and been established by purely human means*, I leave to anyone who wishes to undertake it the discussion of this difficult Problem: which is the more necessary, an already united Society for the institution of Languages, or already invented Languages for the establishment of a Society?'. *SD*, Part I, p. 149 [italics mine].

16 See *SD*, Exordium, pp. 130–3: 'It is of man that I am to speak [...]. O Man, whatever Land you may be from, whatever may be your opinions, listen; Here is your history'.

apes, so iconoclastic and revolutionary[17] that Darwin's work is often reduced to this claim, was neither the starting point nor the core of his intellectual development.

His anthropological questioning emerged late, preceded by a long phase of observation and data collection on plants and animals. Initially destined for the priesthood and a student at Cambridge with no marked scientific vocation, Darwin developed a growing interest in natural history thanks to his encounters with botanist John Stevens Henslow and geologist Adam Sedgwick. This interest deepened during the Beagle expedition (1831–1836), in which he participated as a 'gentleman-naturalist'. The voyage enabled Darwin to observe fossils, island fauna, and coral reef formations, and to return with thousands of samples. It was upon his return that he first formulated, in his 1842 and 1844 *Sketches*, the hypothesis of natural selection as a mechanism of species transformation. He subsequently published several geological and zoological works which established his scientific reputation and refined his skills in comparative anatomy. This long methodological detour exemplifies the empirical rigor that characterizes his entire body of work.

The well-known influence of Malthus came in 1838, when Darwin read the *Essay on the Principle of Population*. From it he drew the insight that species tend to reproduce beyond the availability of resources, and that this 'struggle for existence' creates a selective pressure favoring slightly advantaged individuals. He acknowledges in his *Autobiography* that this was the decisive factor that allowed him to formulate a mechanical explanation of evolution. But this reading could not have had such effects without the groundwork laid

---

17 See Sigmund Freud, *Introductory Lectures on Psycho-Analysis*, Part III: *General Theory of the Neuroses*, in *The Standard Edition of the Complete Psychological Works of Sigmund Freud*, 24 vols, vol. 16, 1916–17, trans. by James Strachey (London: Hogarth Press and the Institute of Psycho-Analysis, 1953), p. 284: 'Humanity has in the course of time had to endure from the hands of science two great outrages upon its naïve self-love. The first was when it realized that our earth was not the centre of the universe [...]. The second was when biological research robbed man of his peculiar privilege of having been specially created, and relegated him to a descent from the animal world'.

by his naturalistic observations. This point is crucial: Darwin never claimed to base his hypotheses on abstract principles. His approach involves a constant back-and-forth between facts, hypotheses, and verification. Indeed, it took twenty-one years for these ideas to be published, in *On the Origin of Species*, and another twelve years before he explicitly addressed the human question in *The Descent of Man*. This delay shows how reluctant Darwin was to extend his theories to humans – unlike Alfred Russel Wallace, a self-taught naturalist and great admirer of Darwin, who had already published an article in 1864 entitled *The Origin of Human Races*.

Yet Wallace diverges from Darwin on a crucial point: he believes that at a certain stage, the human being partially escapes the laws of nature, rising above them by virtue of his intellectual faculties alone. At the beginning of Chapter II of *The Descent of Man*, Darwin expresses surprise that, when it comes to humans, Wallace limits the actual action of natural selection to a very modest improvement in mental capacities, scarcely beyond those of the apes: 'I cannot understand how it is that Mr. Wallace maintains, that "natural selection could only have endowed the savage with a brain a little superior to that of an ape"'. This amounts to saying that, according to Wallace, a large part of human intelligence is not the result of natural selection. He indeed holds that some of the higher human faculties – abstract reasoning, mathematics, art, and morality – are not useful for survival in the primitive state of human beings, and therefore could not have been selected naturally. He thus leans toward a spiritualist or metaphysical (even supernatural) explanation. Darwin's criticism of Wallace is revealing: he refuses to dissociate humans from animals on the basis of speculative hypotheses, even if he acknowledges thresholds and degrees in the evolution of human faculties.

This desire to distinguish himself from philosophical speculation is also apparent in his judgment of Herbert Spencer. In a passage from his *Autobiography* that was not published during his lifetime, Darwin writes:

> Herbert Spencer's conversation seemed to me very interesting, but I did not like him particularly, and did not feel that I could easily have

> become intimate with him. I think that he was extremely egotistical. After reading any of his books, I generally feel enthusiastic admiration for his transcendent talents, and have often wondered whether in the distant future he would rank with such great men as Descartes, Leibnitz, etc., about whom, however, I know very little. Nevertheless I am not conscious of having profited in my own work by Spencer's writings. His deductive manner of treating every subject is wholly opposed to my frame of mind. His conclusions never convince me: and over and over again I have said to myself, after reading one of his discussions, – 'Here would be a fine subject for half-a-dozen years' work'. His fundamental generalisations (which have been compared in importance by some persons with Newton's laws!) – which I daresay may be very valuable under a philosophical point of view, are of such a nature that they do not seem to me to be of any strictly scientific use. […] They do not aid one in predicting what will happen in any particular case.[18]

This passage expresses a distrust of philosophical generalizations and a strong commitment to an empirical definition of scientific validity. Darwin contrasts the possible 'philosophical' value of Spencer's theses with their lack of scientific usefulness, in the sense that they allow neither prediction nor experimentation. He emphasizes that a fruitful hypothesis must be testable – a formulation very close to the criterion of falsifiability that Karl Popper would formalize a century later.[19]

Furthermore, in *The Descent of Man*, he distinguishes, quite significantly, between 'facts' and 'views' ('view', i.e. hypotheses or speculations): 'False facts are highly injurious to the progress of science, for they often endure long; but false views, if supported by

---

18 Charles Darwin, *The Autobiography of Charles Darwin 1809–1882*, with original omissions restored, ed. with Appendix and Notes by his grand-daughter Nora Barlow (New York/London: W. W. Norton, 1958), pp. 108–9.

19 'The criterion of the scientific status of a theory is its falsifiability, or refutability, or testability. […] Confirmations should count only if they are the result of risky predictions; that is to say, if, unenlightened by the theory in question, we should have expected an event which was incompatible with the theory – an event which would have refuted the theory'. Karl Popper, *Conjectures and Refutations: The Growth of Scientific Knowledge* (London: Routledge & Kegan Paul, 1963), p. 36.

some evidence, do little harm, for every one takes a salutary pleasure in proving their falseness; and when this is done, one path towards error is closed and the road to truth is often at the same time opened'.[20]

This primacy of experimental truth over abstract speculation sheds light not only on the difference between Darwin and Spencer, but also between Darwin and Rousseau. Whereas the latter begins with metaphysical or moral hypotheses and 'meditations' to reconstruct a history of humankind, Darwin starts with facts, anatomical and behavioral analogies, from which he deduces transformation mechanisms.

3. *Rousseau, a naturalist?*

Yet the opposition between Rousseau as a philosopher and metaphysician, and Darwin as a scientist, a positivist, and one resistant to anthropological speculation, is somewhat reductive. Rousseau himself constantly maintains a tension between the 'physical' and the 'moral' registers.[21] In the *Second Discourse*, he first contrasts man as a 'physical being' with man considered from a 'metaphysical and moral'[22] point of view – an opposition that precisely introduces the passage on 'perfectibility'. In contrast, in Note X of the same *Discourse*, he refers to a 'natural, moral and political history' of humankind, which places the 'moral' in continuity with the 'natural' – a 'history' that he sees as still to be written, through a scholarly and empirical endeavor whose outline he sketches: it would involve studying humans 'by their conformities and their differences' in order to avoid 'precipitous judgments'.[23] This note reflects a unified approach to human beings, articulating the natural and the moral in a

20 *DM*, Part Two, Chap. XXI, vol. 22, p. 631.

21 This distinction, which is very common among modern philosophers, refers to an opposition between material, sensible realities and mental, possibly immaterial realities.

22 'Until now I have considered only Physical Man; let us now try to view him from the Metaphysical and Moral side'. *SD*, Part I, p. 140.

23 *SD*, Note X, pp. 208 and 210.

comparative perspective.[24] This demand for knowledge is grounded in a practice: travel. For Rousseau, only a genuine experience of concrete human diversity allows one to overcome reductionist conceptions of the human being.[25] This valorization of fieldwork is not without evoking the essential role of Darwin's observations during his voyage on the Beagle.

Even more strikingly, in this same Note X, Rousseau questions the clear boundary between humans and animals. He relies on travelers' accounts of so-called 'anthropomorphic animals' – orangutans, pongos (gorillas or chimpanzees), mandrills – and raises the question of whether these might, in fact, be human beings in a 'savage' state, having failed to develop the potentialities of perfectibility, much like 'feral children'. He mentions the use of fire, the construction of shelters, or even rudimentary burials as ambiguous behaviors. While maintaining perfectibility as the discriminating criterion, he entertains the possibility that certain 'anthropomorphic animals'

---

24 Rousseau is obviously not the only author of his time to consider the possibility of overcoming fixism. Several eighteenth-century thinkers had already begun to challenge the fixity of species and to envision forms of natural transformation. Among them, Georges-Louis Leclerc, Comte de Buffon suggested in his *Histoire naturelle* [1749–88] that species could undergo modifications due to environmental factors, despite his cautious reaffirmation of fixism. Pierre-Louis Moreau de Maupertuis, in *Vénus physique* [1745] and *Essais de cosmologie* [1750], advanced the idea of random variation and survival of the fittest. Denis Diderot, especially in *Le Rêve de d'Alembert* [1769], proposed a materialist and dynamic vision of nature, implying a continuity from matter to life. Jean-Baptiste Robinet, in *De la nature* [1761–8], defended a teleological development of forms culminating in man. While none of these authors provided a systematic evolutionary theory, their ideas contributed to the gradual erosion of classical fixism and paved the way for later transformist doctrines. See on this point Mary Efrosini Gregory, *Evolutionism in Eighteenth-Century French Thought* (Oxford/Bern/Berlin: Peter Lang, 2008).

25 'Let us suppose a Montesquieu, a Buffon, a Diderot, a Duclos, a d'Alembert, a Condillac, or men of that stamp, traveling with a view to instruct their compatriots, [...]; let us suppose that on their return from these memorable travels, these new Hercules set down at leisure the natural, moral and political history of what they had seen, then we would ourselves see a new world issue from their pen'. Ibid., p. 211.

might express rudimentary forms of this faculty, and concedes that the only way to settle the question would be to attempt cross-breeding – an experiment he deems morally unacceptable. In this note, he considers the possibility of an 'intermediate' humanity, that of beings who are perfectible but still 'stupid', in his words, because their perfectibility has not yet yielded visible effects.[26]

Such a hypothesis, although cautiously formulated, introduces the idea of a possible evolutionary continuity between subspecies or at least 'varieties' of man. This perspective is already sketched at the very beginning of the *Second Discourse*, where Rousseau evokes the hypothesis of a man 'in the first embryo of the species, [...] [with] first hooked claws, [...] hairy as a bear and whether, walking on all fours'. He ultimately rejects this hypothesis – which he attributes to Aristotle, though no clear source from Aristotle can be identified[27]– not on principle, but because the natural sciences of his time do not allow it to be settled: 'Comparative Anatomy has as yet made too little progress, the observations of Naturalists are as yet too uncertain to permit establishing the basis of a solid argument on such foundations'.[28] In *Émile*, he supposes that climatic conditions alter natural constitutions, and that peoples living in extreme zones display different cerebral organization: 'It appears [...] that the organization of the brain is less

26 'One finds in the description of these supposed monsters striking conformities with the human species, and smaller differences than might be pointed to between one human being and another. It is not clear from these passages what the Authors' reasons are for refusing to call the Animals in question Savage men, and it is easy to conjecture that it is because of their stupidity, and also because they did not speak; weak reasons for those who know that, although the organ of speech is natural to man, speech itself is nevertheless not natural to him, and who recognize the extent to which his perfectibility may have raised Civil man above his original state. [...] What would have been the judgment of such Observers about the Child found in 1694 [...], who gave no sign of reason, walked on his hands and feet, had no language, and formed sounds in no way resembling those of a man'. Ibid., pp. 205 and 207.

27 This invention of a reference to an imaginary text by Aristotle may be interpreted as a mere approximate citation (Rousseau almost always quoted from memory), but also as a way of making a highly unorthodox hypothesis more acceptable.

28 *SD*, Part I, p. 134.

perfect in the two extremes. Neither the Negroes nor the Laplanders have the sense of the Europeans'.[29] Such remarks reflect an interest in human variation and in a comparative anthropology grounded in empirical data. In addition, several of his writings reveal a keen interest in ethology. He observes animal behavior with curiosity and without prejudice, and he wonders about manifestations of instinct as possible indications of a similarity, at least relative and partial, between animal and human behavior.[30]

4. *From natural selection to sexual selection: the introduction of symbolic mediation in the process of evolution*

Darwin's position, meanwhile, cannot be reduced to strict continuism. While *The Origin of Species* relies on a purely natural

29 *Émile* [1762], Book I, trans. by Christopher Kelly and Eve Grace, in *The Collected Writings of Rousseau*, vol. 9, 2010, p. 179.

30 'Where was it learned that instinct never produces effects in animals similar to those that shame produces in men? I see proofs to the contrary every day. I see some animals hide themselves when satisfying certain needs, in order to keep a disagreeable object from the senses; I see them, instead of fleeing, eager to cover the vestiges afterwards. What is needed for these efforts to have an air of propriety and decency other than that they be taken by men? In their loves I see caprices, choices, and concerted refusals which come very close to following the maxim of exciting the passions by obstacles'. *Letter to D'Alembert* [1758], trans. by Allan Bloom (Ithaca, NY: Cornell University Press, 1960), p. 87. 'Modern philosophy, accepting only what it explains, is careful not to accept that obscure faculty called instinct, which appears without any acquired knowledge to guide animals toward some end. […] I ask what name I ought to give to the ardor with which my dog makes war on moles he does not eat, to the patience with which he sometimes watches for them for whole hours, and to the skill with which he grabs them, throws them out on the earth the moment they push up, and then kills them, only to leave them there, without anyone ever having trained him for this hunt and taught him moles were there? I ask further – and this is more important – why, the first time I threatened this same dog, he lay with his back on the ground, his paws bent back in a supplicant attitude, the one most suited to touch me, a posture he would have certainly not kept if, without letting myself be moved, I had beaten him in this position?'. *Émile*, Book IV, p. 449, note.

mechanism of selection based on two strictly physical and biological variables (random mutations[31] and environmental pressure), *The Descent of Man* introduces a new register: social pressure, judgment, and representations. It is interesting to note that Darwin writes in Chapter IV, just a few lines apart: 'There can be no doubt that the difference between the mind of the lowest man and that of the highest animal is immense', and then 'the difference in mind between man and the higher animals, great as it is, certainly is one of degree and not of kind'.[32]

The shift from a continuist thesis toward a more discontinuist one is supported by observations that are quite close to those Rousseau formulates regarding *perfectibility*:

> An anthropomorphous ape, if he could take a dispassionate view of his own case, would admit that though he could form an artful plan to plunder a garden – though he could use stones for fighting or for breaking open nuts, yet that the thought of fashioning a stone into a tool was quite beyond his scope. Still less, as he would admit, could he follow out a train of metaphysical reasoning, or solve a mathematical problem, or reflect on God, or admire a grand natural scene. Some apes [...] would admit, that though they could make other apes understand by cries some of their perceptions and simpler wants, the notion of expressing definite ideas by definite sounds had never crossed their minds. [...] They would be forced to acknowledge that disinterested love for all living creatures, the most noble attribute of man.[33]

Nevertheless, Darwin never explicitly adopts a discontinuist stance:

> The senses and intuitions, the various emotions and faculties, such as love, memory, attention, curiosity, imitation, reason, etc., of which man boasts, may be found in an incipient, or even sometimes in a well-developed condition, in the lower animals. They are also capable of

---

31 The concept of *mutation* in the modern genetic sense is post-Darwinian. In *On the Origin of Species*, Darwin speaks of 'spontaneous variations' or 'individual differences', having no access to the later discoveries of Mendelian genetics.

32 *DM*, Part One, Chap. IV, vol. 21, p. 130.

33 Ibid.

> some inherited improvement, as we see in the domestic dog compared with the wolf or jackal. If it could be proved that certain high mental powers, such as the formation of general concepts, self-consciousness, etc., were absolutely peculiar to man, which seems extremely doubtful, it is not improbable that these qualities are merely the incidental results of other highly-advanced intellectual faculties.[34]

Darwin therefore assumes that certain complex mental faculties may gradually emerge in other species, and that what is usually considered specifically human might merely be the result of a prolonged development of elementary dispositions.

It is nevertheless necessary to examine more closely how he addresses the question of how complex human dispositions and specific traits can result from evolutionary processes. Regarding social relations, it is well known that he distances himself from Spencer, not only because of the latter's tendency toward speculation, but especially because, unlike Spencer, he refuses to interpret human history solely through the mechanism of competition.[35] The comparative advantage of humans, in his view, lies first and foremost in their capacity to cooperate. He notes that many species that have developed instincts of mutual aid have gained a comparative advantage. With regard to this social instinct, he highlights 'sympathy' as playing a central role, explicitly drawing on the theories of Hume and Bentham. This conception partially aligns with Rousseau's, for whom *pity* constitutes a fundamental moral affect in humans close to the state of nature.[36]

However, Darwin's concept of 'sympathy' is quite complex and cannot be reduced to a form of identification that leads one to suffer from the suffering of others and to enjoy their pleasure. Sympathy is embedded in a social system of value judgments, norms, and

---

34 Ibid.

35 On the fundamental differences between Darwin and Spencer, particularly Darwin's refusal to interpret human history solely in terms of competition, see Richard Weikart, 'Was Darwin or Spencer the Father of Laissez-Faire Social Darwinism?', *Journal of Economic Behavior & Organization*, 71 (2009), 20–8, and H. Allen Orr, 'Darwin and Darwinism: The (Alleged) Social Implications', *Genetics*, 183.3 (2009), 767–72.

36 See *SD*, Part II, pp. 152–3.

preferences. It is the judgments rendered by others – praise or blame – that guide behaviors. The major evolutionary criterion thus becomes no longer the survival of the fittest, but the success of the most admired, the most loved, the most respected:

> In order that primeval men, or the ape-like progenitors of man, should become social, they must have acquired the same instinctive feelings, which impel other animals to live in a body; and they no doubt exhibited the same general disposition. They would have felt uneasy when separated from their comrades, for whom they would have felt some degree of love; they would have warned each other of danger, and have given mutual aid in attack or defence. All this implies some degree of sympathy, fidelity, and courage. Such social qualities, the paramount importance of which to the lower animals is disputed by no one, were no doubt acquired by the progenitors of man in a similar manner, namely, through Natural Selection, aided by inherited habit. [...]
>
> But another and much more powerful stimulus to the development of the social virtues, is afforded by the praise and the blame of our fellow-men. To the instinct of sympathy [...], it is primarily due, that we habitually bestow both praise and blame on others, whilst we love the former and dread the latter when applied to ourselves; and this instinct no doubt was originally acquired, like all the other social instincts, through Natural Selection.[37]

In other words, natural selection may have genetically favored another selective mechanism: that of social recognition. Behaviors are no longer advantaged simply because they enhance survival, but because they are appreciated by others. The human environment, as a structure of normative judgments, becomes the primary factor of selection. This is why Darwin chooses to speak not of 'natural selection' but of 'sexual selection'. He particularly emphasizes the importance of social power in access to sexual partners: wealth, hierarchical status, possession of goods (animals, weapons), and seduction skills—all of these factors determine reproductive success.[38]

37 *DM*, Part One, Chap. IV, vol. 21, pp. 134–6.
38 *DM*, Part Two, Chap. XX, vol 22, p. 619.

The role of aesthetic judgment in the choices leading individuals to prefer certain mates is also extensively developed. Darwin even considers it a major factor in the diversification of the human species into 'races', based on phenotypic criteria. He provides many examples of physical traits selected because they align with norms of beauty as valued within a given group:

> Mr. Winwood Reade informs me that the Jollofs, a tribe of negroes on the west coast of Africa, 'are remarkable for their uniformly fine appearance'. A friend of his asked one of these men, 'How is it that every one whom I meet is so fine looking, not only your men but your women?' The Jollof answered, 'It is very easily explained: it has always been our custom to pick out our worst-looking slaves and to sell them'. It need hardly be added that with all savages, female slaves serve as concubines. That this negro should have attributed, whether rightly or wrongly, the fine appearance of his tribe to the long-continued elimination of the ugly women is not so surprising as it may at first appear.
>
> [...]
>
> Each race has its own style of beauty [...]. If then the several foregoing propositions be admitted, and I cannot see that they are doubtful, it would be an inexplicable circumstance if the selection of the more attractive women by the more powerful men of each tribe, who would rear on an average a greater number of children, did not after the lapse of many generations somewhat modify the character of the tribe.[39]

Darwin stresses an important point: in many cases, these selection criteria do not confer any objective adaptive advantage. Certain physical variations have been favored or eliminated not through natural necessity but through the power of social approval or disapproval. This is particularly evident and striking in the case of hairiness:

> From the presence of the woolly hair or lanugo on the human foetus, and of rudimentary hairs scattered over the body during maturity, we may infer that man is descended from some animal which was born hairy and remained so during life. The loss of hair is an inconvenience

39 Ibid., pp. 611 and 620.

> and probably an injury to man, even in a hot climate, for he is thus exposed to the scorching of the sun, and to sudden chills, especially during wet weather. […] The natives in all countries are glad to protect their naked backs and shoulders with some slight covering. No one supposes that the nakedness of the skin is any direct advantage to man; his body therefore cannot have been divested of hair through Natural Selection.[40] […]
>
> The absence of hair on the body is to a certain extent a secondary sexual character; for in all parts of the world women are less hairy than men. Therefore we may reasonably suspect that this character has been gained through Sexual Selection.[41]

On this subject, Darwin draws on popular sayings and the observation of cultural practices: 'With respect to the completion of the process through Sexual Selection, it is well to bear in mind the New Zealand proverb, "There is no woman for a hairy man". […] We know that with savages the men of the beardless races take infinite pains in eradicating every hair from their faces as something odious'.[42] A lack of body hair is considered unattractive in certain cases and on certain parts of the body, while an abundance of hair is considered attractive in other cases. Darwin notes the importance of the beard in men, both as an inherited ornament and as a factor of selection: 'Man in all probability owes his beard […] to inheritance from an ancient progenitor who thus gained his ornaments'.[43]

Individuals would therefore regulate their behaviors based on judgments relating to appearances or behaviors perceived as admirable. Those most admired would consequently gain an advantage in terms of inclusion, prestige, and reproduction. This shift in the evolutionary center of gravity – from *natural* selection to *social* selection – is crucial.[44] This selection process gains

---

40 See Charles Darwin, *Contributions to the Theory of Natural Selection* (London & New York: Macmillan & Co, 1870), p. 346. [Darwin's note].

41 *DM*, Part Two, Chap. XX, vol 22, p. 624–5.

42 Ibid., pp. 625 and 627.

43 Ibid., p. 622.

44 The major and monumental, but also controversial, work of Patrick Tort offers fascinating insights on this issue, primarily through the notion of

additional philosophical significance insofar as Darwin derives from it a genealogy of moral norms, partially grounded in intersubjective relationships.

5. *Two moralists?*

This phenomenon of comparative preference is indeed found in Darwin's explanations when he examines the origin of moral standards. Chapter IV of *The Descent of Man* is devoted to what might be called, borrowing Nietzsche's expression, a 'genealogy of morals'. In this chapter, Darwin seems to oscillate – without fully resolving the tension – between two conceptions of moral duty: on the one hand, a purely naturalistic account, where 'duty' is essentially the same whether it applies to the moral conscience of a human being or to the behavior of a well-trained hunting dog; on the other hand, a normative conception, which distinguishes between superior and inferior forms of morality. In this second view, Darwin explicitly criticizes customary, superstitious, or socially conditioned forms of morality, in contrast with a truly rational and universal moral sense.

At the beginning of Chapter IV, he writes: 'It was assumed formerly by philosophers that the foundation of morality lay in a form of Selfishness', explicitly to refute this thesis. He quotes with admiration Kant: 'duty! thou sublime and mighty name who dost embrace nothing charming or insinuating but requirest submission'. He also criticizes modern utilitarianism: 'It is more correct to speak of the latter principle [the 'greatest happiness principle'] as the standard, and not as the motive of conduct'.[45] According to him, moralists are mistaken in believing that every act must be motivated by some form of pleasure or displeasure. One must introduce a

the 'reversive effect of evolution'. It is impossible, within the scope of this article, to elaborate on or discuss these works. See in particular *Darwin et la science de l'évolution* (Paris: Gallimard, 2000), *Darwin and the Science of Evolution* (New York: Harry N. Abrams, 2001), trans. by Paul G. Bahn; *L'effet Darwin. Sélection naturelle et naissance de la civilisation* (Paris: Seuil, 2008).

45 *DM*, Part One, Chap. IV, vol 21, pp. 101 and 124.

factor too often overlooked in order to understand why unpleasant, dangerous, or even self-sacrificial behaviors may arise spontaneously, even among animals: conditioning and habit. He mentions Mill as one of the few philosophers who understood that certain actions may result from habit alone, 'without the anticipation of pleasure'.[46] For Darwin, 'A belief constantly inculcated during the early years of life [...] appears to acquire almost the nature of an instinct', because 'the very essence of an instinct is that it is followed independently of reason'.[47] This explains how a person may act impulsively against their own interest, 'without any consciousness of pleasure'.[48] He gives the example of a man throwing himself into a fire to save someone else: 'an impulsive power widely different from a search after pleasure or happiness'.[49] This moral impulse derives, in his view, from social instincts inherited from our animal ancestors. He defines their purpose not as happiness, but as 'the general good', that is, 'the rearing of the greatest number of individuals in full vigour and health, with all their faculties perfect'.[50] It is repetition that enables man to master his initial impulses: 'Man, prompted by his conscience, will through long habit acquire such perfect self-command [...] that his desires and passions will at last yield instantly and without a struggle'.[51]

Obedience to a morally perfect duty can therefore take the form of a quasi-instinctive reflex. It is in this context that Darwin refers to the behavior of hunting dogs: 'We even say that a pointer ought to point, and a retriever to retrieve game. If they fail to do so, they fail in their duty and act wrongly'.[52] In this sense, the word 'ought' does not primarily designate an abstract moral law, but rather the expected conformity to behaviors that have been learned and selected.

But in the case of human beings, this 'duty', shaped by habit and conditioning, is directly linked to the opinion of others in the social

46 Ibid., p. 124, note.
47 Ibid., p. 126.
48 Ibid., p. 124.
49 Ibid., p. 125.
50 Ibid.
51 Ibid., p. 119.
52 Ibid.

group – 'the judgment of the community'. This mediation through collective judgment and intersubjectivity introduces a mechanism that is qualitatively different from that of purely physical natural selection. Yet Darwin does not explain how or why certain ancestors of humans and great apes began to act on the basis of values and normative judgments. While it is certainly not inconceivable that such evolution could result from natural selection alone, the very nature of the evolutionary process is fundamentally transformed by the theses Darwin advances in *The Descent of Man*. Moreover – and this is not the least of the paradoxes in what he develops specifically in this work – he criticizes the creation, through this very process, of social norms that he considers absurd and irrational:

> Merely to know that our friends and equals despise us […] is enough to cause great misery. Who can doubt that the refusal to fight a duel through fear has caused many men an agony of shame? […] The breach of a rule held sacred by the tribe, will thus, as it seems, give rise to the deepest feelings […]. How so many strange superstitions have arisen throughout the world we know not. […] How so many absurd rules of conduct, as well as so many absurd religious beliefs, have originated, we do not know; nor how it is that they have become, in all quarters of the world, so deeply impressed on the mind of men […] Neither can we say why certain admirable virtues, such as the love of truth, are much more highly appreciated by some savage tribes than by others.[53]

Darwin thus refers to the 'Law of Honour', whose transgression may cause 'more agony than a real crime'.[54] A Hindu who violates the dietary restrictions of his caste may feel more remorse than if he had committed theft; an Australian Aboriginal may be tormented by the customary obligation to kill a woman as an act of expiation. Here we understand the importance of examples relating to aesthetic judgments in criteria of preference, particularly sexual preference. Since aesthetic taste is subjective and lacks any natural or rational foundation, it is not surprising that the selection of moral behaviors deemed honorable or admirable is equally arbitrary and varies from one social group to another.

53 Ibid., pp. 118–9 and 126.
54 Ibid., pp. 125–6.

Yet, unlike Nietzsche, Darwin does not draw relativistic conclusions from this. He considers it legitimate to evaluate moral systems according to universal ethical principles. He praises values and ethical behaviors based on self-mastery, reason, and asceticism. This is illustrated by a quotation from *The Idylls of the King* by Tennyson,[55] which evokes the control of thought over itself: ‘not even in inmost thought to think again the sins that made the past so pleasant to us’. On this point, Darwin also cites, as noted above, Emmanuel Kant, as well as Herbert Spencer, whom he refers to at that moment as a ‘great philosopher’: ‘The experiences of utility organized and consolidated through all past generations [...] have become in us certain faculties of moral intuition’, and Marcus Aurelius: ‘Such as are thy habitual thoughts, such also will be the character of thy mind; for the soul is dyed by the thoughts’.[56]

Darwin thereby makes rational self-mastery – the ideal of ancient Stoicism – the culmination of moral culture, and even considers it both possible and desirable that genuinely virtuous habits might become hereditary, ensuring the gradual triumph of virtue over selfish desires. He therefore articulates three levels of ‘morality’: (1) one based on immediate instinct; (2) another based on conditioning and habit – which may be irrational and ethically questionable, since arbitrary –; and (3) a final one based on self-mastery and the internalization of rules that are objectively conducive to the ‘general good’. In this complex evolutionary model, human morality begins with conditioned reflex and culminates in reflective control.

It is particularly interesting to return here to Rousseau, as he too proposes a heteronomous genealogy of morality, grounded in judgments of preference, concern for ‘blame’ and ‘praise’, and

55 *Idylls of the King*, published between 1859 and 1885, is a cycle of twelve narrative poems which retells the legend of King Arthur.

56 *DM*, Part One, Chap. IV, vol 21, p. 127. The quotation from Tennyson is taken from the *Merlin and Vivien* section of *The Idylls of the King*. The quotation from Herbert Spencer comes from a *Letter to Mr. Mill*, cited in Alexander Bain's *Mental and Moral Science* (London: Longman, Green & Co., 1868), p. 722. The quotation from Marcus Aurelius is taken from *The Thoughts of the Emperor M. Aurelius Antoninus*, English trans., 2nd ed. (London: George Bell & Sons, 1869), p. 112.

choices in love. This can lead to arbitrary and ethically questionable behaviors, since they are driven by *amour-propre* and public opinion—but also, under certain conditions, to genuine ethical improvement. Rousseau distinguishes a deleterious version of self-love, associated with the term 'vanity'.[57] But he also identifies a legitimate use of self-love, one that expands the soul rather than diminishing it. The *Second Discourse* stresses the link between self-love and the feeling of 'honour'.[58] Honour is the desire to obtain the esteem of others and can thus lead individuals to offer the best of themselves. Several passages in *Émile* underscore this ambivalence of self-love. Self-love is 'pride in great souls, vanity in small ones'. What Rousseau writes about love and friendship aligns with this perspective:

> True love [...] will always be honored by men; for, although its transports lead us astray, although it does not exclude odious qualities from the heart that feels it – and even produces them – it nevertheless always presupposes estimable qualities without which one would not be in a condition to feel it. [...] One wants to obtain the preference that one grants. Love must be reciprocal. To be loved, one has to make oneself lovable. To be preferred, one has to make oneself more lovable than another, more lovable than every other, at least in the eyes of the beloved object.[59]

Preferential love is indeed a modality of self-love since it compares, ranks, and aims to be preferred by those whom one prefers.[60] But unlike domineering self-love, which diminishes others in order to exalt oneself, preferential love contains a form of 'remedy in the evil itself'.[61] If the lover wishes to be preferred, it is not to be envied

57 *Émile*, Book IV, p. 494.

58 *SD*, Note XV, p. 219.

59 *Émile*, Book IV, pp. 364–5. Cf. ibid., Book V, pp. 617 and 798; *Julie, or the New Heloise*, Book III, Letter 7, ed. and trans. by Philip Stewart and Jean Vaché, in *The Collected Writings of Rousseau*, vol. 6, 1997, pp. 262–3.

60 See *SD*, Part I, p. 155.

61 The expression 'remedy in the evil itself' ['remède dans le mal'] appears in *The Confessions*, Book I, trans. by Christopher Kelly, in *The Collected Writings of Rousseau*, vol. 5, 1995, p. 14. See Jean Starobinski, *Blessings*

or feared, but to be loved. And the desire to be preferred implies the obligation to be lovable, to deserve the esteem of the loved one. This demand can lead to moral elevation in the lover's behavior. A lasting romantic relationship is impossible if the lovers do not strive to merit each other's regard, which means cultivating what makes them worthy of that regard. The great romantic passions described by Rousseau contain this dimension of moral elevation. The greatest and almost only pleasure of Julie and Saint-Preux is to be virtuous, to esteem each other, and to encourage each other to virtue.

What is true of love also applies to friendship. The man who seeks to be esteemed aims to find his rightful place in the relationship, and reciprocally, he helps the person he loves to judge himself rightly. The friend is morally demanding – and has the right to be so, for this demand is reciprocal. Rousseau writes of the adult Émile: 'He loves men because they are his fellows, but he will especially love those who resemble him most because he will feel that he is good [...]. He will [...] say to himself, [...] "I rejoice because they approve of what I have done that is good. I rejoice that the people who honor me do themselves honor. So long as they judge so soundly, it will be a fine thing to obtain their esteem"'.[62]

6. *Conclusion*

The German philosopher Friedrich Albert Lange interprets Darwin's legacy as an invitation to rethink the place of man in nature without reducing ethics to its biological determinations. According to him, 'The moral law, instead of being a special revelation, becomes a natural product of mankind, and this again is the result of the social instincts and the struggle for existence'.[63] As Chiara Russo

in Disguise, or The Morality of Evil, trans. by Arthur Goldhammer (Cambridge, MA: Harvard University Press, 1993). Originally published as *Le remède dans le mal : critique et légitimation de l'artifice à l'âge des Lumières* (Paris: Gallimard, 1989).

62 *Émile*, Book IV, p. 511.

63 Friedrich A. Lange, *History of Materialism*, 2 vols, vol. 2, trans. by Ernest C. Thomas (London: Trübner & Co, 1877), p. 304.

Krauss shows, Lange interprets Darwinism in light of Schillerian aesthetic Kantianism: the ethical ideal would first derive from an 'aesthetic redemption' in two stages – first, producing an image of ideal beauty; then, realizing it in social reality – so that the aesthetic Idea and the moral Ideal are linked as dialectical moments in human evolution.[64] Here we find a striking example of the continuum that connects natural demands to moral ones, via the judgment of admiration previously discussed in Rousseau and Darwin.

More generally, one must probably turn to Kant to understand why it remains impossible to choose dogmatically between a purely materialist conception of man and a dualistic one asserting the existence of an immaterial soul. Nothing illustrates this better than the brilliant *Antinomy of Pure Reason* from Book II of the *Critique of Pure Reason*. In the 'Second Conflict' of this Antinomy, the Thesis reads: 'Causality in accordance with laws of nature is not the only one from which all the appearances of the world can be derived. It is also necessary to assume another causality through freedom in order to explain them'. The Antithesis reads: 'There is no freedom, but everything in the world happens solely in accordance with laws of nature'. Kant summarizes the meaning of this antinomy as follows:

> The question of freedom of the will which has always put speculative reason into such embarrassment is really only transcendental, and it concerns only whether a faculty of beginning a series of successive things or states from itself is to be assumed. How such a faculty is possible is not so necessary to answer, since with the existing laws in accordance with natural laws we likewise have to be satisfied with the a priori cognition that such a thing must be presupposed, even though we do not in any way comprehend how it is possible.[65]

64 See Chiara Russo Krauss, 'The Struggle for Existence and the Ideal: Friedrich Albert Lange's Political Teleology at the Intersection of Schiller and Darwin'. *Geltung. Journal of Studies on the Origins of Contemporary Philosophy*, 1.2 (2021), 1–74.

65 Immanuel Kant, *Critique of Pure Reason*, Book II, *The dialectical inferences of pure reason*, Chapter II, *The antinomy of pure reason, Second conflict*, ed. and trans. by Paul Guyer and Allen W. Wood (Cambridge, UK: Cambridge University Press, 1998), pp. 484–6.

Thinking the possibility of freedom, particularly moral freedom, is necessary. *Knowing* the empirical conditions of this freedom is impossible. Rousseau had already grasped perfectly, in the *Second Discourse*, this difference between what philosophy must postulate and what empirical knowledge can demonstrate: this theoretical tension is, as we have seen, one of the reasons for the elaboration of his theory of perfectibility. The qualitative differences in the modalities of evolution, as theorized by Darwin, depending on whether we are dealing with natural selection or sexual selection, confirm the immense difficulty of adopting a strictly physiological and materialist perspective to understand the process of hominization. The necessity of passing through the mediation of mental representations and value judgments – what Cassirer calls a 'system of symbolic forms'[66] – certainly does not prove the spirituality of the soul, but bears witness to the specific and irreducible problems posed by the knowledge of man by man.

66 Ernst Cassirer, *The Philosophy of Symbolic Forms*, 3 vols, Vol. 2: *Mythical Thought*, trans. by Ralph Manheim (New Haven: Yale University Press, 1955), Introduction, p. 21.

Giulia Abbadessa*

LEOPARDI'S SCEPTICAL HUMANISM BETWEEN DARWINISM AND NIHILISM

Abstract

This article explores the humanistic dimension of Leopardi's philosophy through an analysis of the *Zibaldone*, the *Operette Morali*, and the *Pensieri*. The aim is to highlight how Leopardi progressively rethought the relationship between humans and animals, celebrating and radically renewing the humanistic tradition from a sceptical and ironic perspective. This approach reaches its peak in the *Pensieri*, where Leopardi offers a nihilistic reinterpretation of Pico della Mirandola's *Discorso sulla dignità dell'uomo*.

Keywords

Humanism – Scepticism – Animality – Boredom – Leopardi

1. *Introduction*

Critics have recently returned to questioning Leopardi's philosophy of nature, offering the first modern edition of his *Compendio di storia naturale* (1812) and investigating his in-depth knowledge of the scientific culture of the time, starting from the sources recalled in his youthful writings, such as *Le spectacle de la nature* by Noël-Antoine Pluche or the *Fondamenti della scienza chimica-fisica applicate al studio dei corpi e alle forze della natura* by Vincenzo Dandolo,[1] up

---

* Università degli Studi di Napoli Federico II.

1 Cf. Gaspare Polizzi, *Corporeità e natura in Leopardi* (Milan–Udine: Mimesis, 2023), Chiara Fenoglio (ed.), *Un mondo 'inconcepibile'. Saggi su Leopardi e la Natura* (Venice: Marsilio, 2024) and Tatiana Crivelli, Patrizia

to the proto-evolutionist ones, among which also figures the work of Edward Darwin, grandfather of Charles Darwin.[2] If the philosophy of nature constitutes the *trait-d'union* between Leopardi's speculative philosophy and practical philosophy,[3] recent criticism has also advanced the possibility of interpreting the thought of the poet from Recanati in terms of an eco-pessimism,[4] rethinking the anti-humanist or post-human dimension of his work.[5] Re-examining the connection between Leopardi's philosophy and Darwinism, this essay instead analyses the *Zibaldone*, the *Operette morali* but also the *Pensieri*, in order to highlight their more specifically humanist dimension.

2. *Leopardi's Humanism between Natural Philosophy and Darwinism*

After having meditated on the difference between the animal and the human since his early youth, with the fable *I Filosofi e il Cane* (1810) and, for an exercise related to his Catholic education, with the *Dissertazione sopra l'anima delle bestie* (1811), Leopardi continued to reflect on this theme in the *Zibaldone* in which the poet aimed to elaborate a system of thought[6] that he himself defined in 1821

Landi (ed.), 'Giacomo Leopardi e le rappresentazioni sette-ottocentesche della natura', *Rivista internazionale di studi leopardiani*, 13 (2020).

2 Anna Clara Bova, *Leopardi e la zoonomia di Erasmo Darwin*, in *La prospettiva antropologica nel pensiero e nella poesia di Giacomo Leopardi*, Atti del XII Convegno internazionale di studi leopardiani, Recanati, September 23–26, 2008, ed. by Chiara Gaiardoni (Florence: Olschki, 2010), pp. 477–91.

3 Gaspare Polizzi, *'...per le forze eterne della materia'. Natura e scienza in Giacomo Leopardi* (Milan: FrancoAngeli, 2008).

4 Franco D'Intino (ed.), '"Eco-Leopardi". Visioni apocalittiche e critica dell'umano nel poeta della natura', *Costellazioni*, 10 (2019).

5 Valerio Camarotto, Laura Fournier-Finocchiaro and Martina Piperno (ed.), 'Technique, animalité et post-humanisme chez Giacomo Leopardi / Tecnica, animalità e postumanesimo in Giacomo Leopardi / Technology, Animality, and Posthumanism in Giacomo Leopardi', *Cahiers d'études italiennes*, 40 (2025).

6 Giacomo Leopardi, *Zibaldone di pensieri*, ed. by Lucio Felici, Emanuele Trevi, Marco Dondero and Wanda Marra (Rome: Newton Compton,

as a 'scetticismo ragionato e dimostrato'.[7] Due to its fragmentary style, moreover, the *Zibaldone* follows the tradition inaugurated by Montaigne's *Essais* which, as is known, was renewed in France between the seventeenth and eighteenth centuries by various authors who equally inspired Leopardi, including Pascal, Voltaire and Rousseau, all of whom identified a link between the animal and the human, like other exponents of libertine and Enlightenment culture such as Gassendi and Condillac, La Mettrie and d'Holbach, equally well known to the poet of Recanati.[8]

In the *Zibaldone*, Leopardi took up both the more autobiographical dimension of the *Essais* and Montaigne's favored philological model of citing numerous ancient and modern authors, thus establishing a dense intertextual dialogue with the Republic of Letters. Like Montaigne, Leopardi therefore adopted a humanist research method, founded, as Francesco Tateo points out, on rewriting as interpretation,[9] or rather, on the renewal of models conceived, in the words of Marc Fumaroli, as flowers from which to extract, like bees, a new honey.[10]

Although he did not adopt an extensive and explicit system of quotations like that of Montaigne and Leopardi, like the latter, Charles Darwin not only paid great attention to etymologies and

2001) [from now *Zib.*], p. 946 (April 16–17, 1821): 'Qualunque pensatore, e i piú grandi massimamente, hanno avuto ciascuno il loro sistema [...]. Cartesio, Malebranche, Newton, Leibnizio, Locke, Rousseau, Cabanis, Tracy, de Vico, Kant'. For the translation see G. Leopardi, *Zibaldone*, ed. by Michael Caesar and Franco D'Intino, trans. by Kathleen Baldwin et alii (New York: Ferrar, Straus and Giroux, 2013).

7 *Zib.*, p. 1655 (September 8, 1821).

8 After having planned in 1828 to write some 'Saggi, alla Montaigne', in 1836 Leopardi acknowledged that he had only written 'des essais en comptant toujours préluder': the quotations are studied by Vincenzo Allegrini, *Un disegno che 'se ne andrà col vento'. Leopardi 'in compagnia' di Montaigne*, in *Oltre le righe. Usi e infrazioni dello spazio testuale*, ed. by Vincenzo Allegrini, Sara De Simone, Alessandra Forte and Dario Panno-Pecoraro (Pisa: Edizioni della Normale, 2020), pp. 196–216.

9 Francesco Tateo, *Riscrittura come interpretazione. Dagli umanisti a Leopardi* (Rome–Bari: Laterza, 2001), pp. 95–166.

10 Marc Fumaroli, *La Querelle des anciens et des modernes: 17–18 siècles*, preceded by *Les abeilles et les araignées* (Paris: Gallimard 2001).

language,[11] but also inscribed his own work in the wake of the research that preceded him,[12] drawing great inspiration from literary and poetic language:[13] emblematic in this sense is the echo detected by critics in his work of *The Botanic Garden* by his grandfather Erasmus Darwin and of Milton's *Paradise Lost*, which were two sources also for Leopardi.[14] Like the poet-philosopher, Charles Darwin can be considered in this sense as an another exponent of humanism, if this tradition, as Ernesto Grassi observed when studying the Renaissance, is based on the dialogue with the word of the past, or on the task of mediating it primarily through the study of metaphorical and metamorphic language, typical of the ancients and aimed at accounting for a phenomenological richness rather than an abstract metaphysics of beings.[15] Compared to the Renaissance humanist tradition, however, it is known that Darwin and Leopardi preferred a scientific use of metaphorical language, striving to grasp the connections between particulars in a system, in the belief that analogical thought consists in the search for organic relationships.[16]

---

11 Cf. Stefano Gensini, 'Darwin's View of Language in The Descent of Man: An Intertextual Reading', *Human Evolution*, 29.4 (2014), 303–18.

12 Cf. Antonella Anedda, *Le piante di Darwin e i topi di Leopardi* (Novara: Interlinea, 2022).

13 Cf. Roger M. White, Jonathan Hodge, Gregory Radick (ed.), *Darwin's Use of Metaphor in the Origin*, in *Darwin's Argument by Analogy: From Artificial to Natural Selection* (Cambridge: Cambridge University Press, 2021), pp. 160-81.

14 Cf. John Holmes, 'Epic Poetry and the Origins of Evolutionary Theory', *Romanticism and Victorianism on the Net*, 66-67 (2016), https://www.erudit.org/en/journals/ravon/2016-n66-67-ravon05340/1069961ar/, and Giulia Abbadessa, 'La letteratura inglese e Leopardi: Milton, Shakespeare, Byron', *La Cultura*, 3 (2021), pp. 443–68.

15 Ernesto Grassi, *I primi scritti 1922–1946* (Naples: La Città del Sole, 2011), p. 881. On Darwin's humanism see Telmo Pievani, *La rivoluzione darwiniana e la scienza dell'uomo*, in *Conversazioni sull'Origine dell'uomo 150 anni dopo Darwin*, ed. by Fabio Salomone and Flavia Di Vincenzo (Monte Compatri: Edizioni Espera, 2021), pp. 31–9.

16 *Zib.*, pp. 946–9 (April 16, 1821). Cf. Anna Clara Bova, *Al di qua dell'infinito. La 'teoria dell'uomo' di Giacomo Leopardi* (Rome: Carocci, 2009) and Anedda, *Le piante*, pp. 140–1.

Although Eugenio Garin has partly distanced himself from Grassi's reading of humanism 'as a philosophy of language',[17] highlighting instead its civil and anthropological dimension, this dimension too is equally present in the work of Darwin as in that of Leopardi who, as Antonella Anedda has points out, all studied nature also to criticize anthropocentrism and slavery, presenting man as an animal—that is, not as the top of the hierarchy of species, but rather as an integral part of it, marked by a great capability to adapt and tragically characterized above all by violence.

3. *For a human-animal society*

In his creative work Leopardi anchored his philosophy of nature primarily in ancient tradition, probably to avoid the censorship that affected his writings – such as the *Paralipomeni della Bratacomiomachia* – also because of his vision of the animal world.[18] Conversely, in the private writing of the *Zibaldone* Leopardi was able to cite even forbidden authors and meditate more freely on the 'differenza di vita fra le bestie e l'uomo', interpreting it in 'accidentali' terms and abandoning, like Darwin, a theological perspective: the poet initially advanced the hypothesis that this difference depended on the 'diversa conformazione del corpo umano più atta alla società'.[19] In 1820, however, in the wake of a philosophical tradition deeply rooted in the ancient world – from Plutarch to Seneca, through Virgil's *Georgics*, in which bees are a counter-model to human society – Leopardi observed that 'la società non è già propria del solo uomo. Le formiche la fanno per

17 Eugenio Garin, *L'Umanesimo italiano* (Rome–Bari: Laterza, 2025), p. XVI. Nonetheless, as Garin himself underlined, an anthropological and civil perspective is also present in Grassi's studies, which insists on the pedagogical purpose of Renaissance linguistic philosophy.

18 Cf. Antonio Giuliano, *Giacomo Leopardi e la Restaurazione: Nuovi documenti* (Naples: Accademia di Archeologia, Lettere e Belle Arti, 1998).

19 *Zib.*, p. 56.

trasportar pesi. Le api hanno anche un governo'.[20] Along with ancient sceptical thought but also with modern, libertine, and Enlightenment philosophy, Leopardi therefore hypothesized that animals possessed 'il principio di ragionamento', specifying that 'qualunque essere non è macchina, ha bisogno di credenze per vivere'.[21] After all, beginning with the fable *I Filosofi, e il Cane*, Leopardi had already begun to explore animal voice in imaginative ways, fantasizing in one of the first pages of the *Zibaldone* about the language shared by animals, drawing inspiration from the Renaissance tradition, and in particular from Sannazzaro's *Arcadia*.[22]

In his philosophical diary, reflecting on social institutions, the poet also observed that 'la natura ha dettato' a few laws, including *in primis* that—already mentioned by Montaigne—of the care of the 'figliuolanza' but also the 'legge di matrimonio' which, although cited by Horace ('*concubitu prohibere vago*'), Leopardi noted was also widespread among the birds that 'si fabbricano il talamo', therefore considering it necessary only 'finché bisogna all'educazione sufficiente dei prodotti di quel tal matrimonio': *vice versa*, 'aritifizi*ale*' is the society that 'necessita la perpetuità de' matrimoni, e la distinzione delle famiglie e delle possidenze'.[23] After criticizing the classical view of society, Leopardi finally concluded that that of ants, aimed at the 'comun bene', is more natural than 'una società più stretta e formata' like the human one, which, instead, lacks unity:

> Così dunque la società veramente primordiale, e naturale alla specie umana, come a quelle dei bruti, senza principato, senza soggezione, senza disuguaglianza, senza gradi, senza regole, poteva benissimo corrispondere al fine, cioè al comun bene, come vi corrisponde quella delle formiche: al qual fine non può mai corrispondere una società piú stretta e formata, se manca di unità. Donnola, castori e gru formano società intere costanti.[24]

20 *Zib.*, p. 210 (August 14, 1820).
21 *Zib.*, p. 437 (December 22, 1820).
22 *Zib.*, p. 55.
23 *Zib.*, pp. 249–50 (September 9, 1820).
24 *Zib.*, p. 587 (January 29–31, 1821).

Since the *Discorso di un italiano intorno alla poesia romantica* (1818) Leopardi had praised nature over reason and civilization. In the *Zibaldone* he valued the animal world above all from a political point of view: on one hand the adopted the anarchic perspective that emerges from the above-mentioned passage; on the other, in line with the Italian philosophy of the Renaissance, he saw in nature a model for the 'condizione umana'[25] and its pursuit of happiness—obscured by the myth of progress and its rationalistic ideologies.

> Dice Machiavelli che a voler conservare un regno, una repubblica o una setta, è necessario ritirarli spesso verso i loro principii. Cosí tutti i politici. Vedi Montesquieu, *Grandeur* ec. ch. 8. [...]. Ed io dico nello stesso senso: a voler conservare gli uomini, cioè farli felici, bisogna richiamarli ai loro principii, vale a dire alla natura. – Oh pazzia. Tu non sai che la perfettibilità dell'uomo è dimostrata. — Io vedo che di tutte le altre opere della natura è dimostrato tutto l'opposto, cioè che non si possono perfezionare, ma alterandole si può solamente corromperle, e questo principalmente per nostra mano.[26]

Leopardi developed his political philosophy from a naturalistic and vitalistic perspective, also considering, on the one hand, in harmony with Voltaire, Rousseau, and Montesquieu, love of country as a virtuous form of self-love, stimulated by nature;[27] and on the other, in tune with Machiavelli before the Enlightenment thinkers, studying the decay that accompanies the evolution of forms of government as a natural process. It is well known that, in Leopardi's political vision—which also drew a parallel between the end of the Roman Republic and the defeat of the French Revolution—the most ancient form of government, namely the primitive monarchy, is succeeded by the republic, which is, however, destined to degenerate with the development of reason, which transforms virtue into a

25 *Zib.*, p. 66.

26 *Zib.*, p. 222 (August 23, 1820).

27 See G. Abbadessa, *La satira di Leopardi contro l'Ottimismo di Leibniz. Per un'ironia più che romantica*, in *Lebenskunst nach Leopardi. Anti pessimistische Strategie im Werk Giacomo Leopardi*, ed. by Milan Herold and Barbara Kuhn (Tübingen: Narr Francke, 2020), pp. 201–28.

phantom, producing tyranny and selfishness, at least until illusions and natural passions resurface.[28]

At the same time, Leopardi developed his natural philosophy from a perspective not only political but also aesthetic, for example by defining as extremely pleasant the sight of both a multitude of small-scale natural elements and the fury of larger ones, such as the sight 'di una folla, o di un gran numero di formiche, o del mare agitato'.[29] Leopardi thus revisited the theory of the sublime, popular in the eighteenth century thanks to Burke, from an anthropological perspective, in tune with his vision of nature as a force capable of countering the immobility and boredom of civilization. While it is not possible to delve into Leopardi's vision of the sublime here, it is worth noting that it was in 1823 that the poet openly formulated a vision of epistemology in which aesthetics and ethics coincided: in the wake of his humanistic studies on primitive imaginative language, already praised by Vico, Leopardi recognized that nature is entirely focused on a poetic purpose, impossible to grasp with a purely rational and analytical approach, which reduces nature itself to a 'corpo morto'.[30]

Although Leopardi had opposed nature to reason since his youth, he progressively demystified its myth, made famous at the time by Rousseau. It's needless to recall that, especially from July 1820 onward, Leopardi developed his famous theory of pleasure, revisiting Pascal's *Pensées* in considering an infinite, material desire for happiness, impossible to satisfy, inherent in the 'anima umana' and in all 'viventi'.[31] In line with this tragic vision, culminating in the recognition of a 'contraddizione spaventevole',[32] through the technique of personification already employed by Montaigne

28 *Zib.* pp. 543–80 (January 22–29, 1821), cf. Sebastiano Timpanaro, *Classicismo e Illuminismo nell'Ottocento italiano* (Pisa: Nistri–Lischi, 1977), p. 150.

29 *Zib.*, p. 1747 (September 20, 1821).

30 *Zib.*, pp. 3237–45 (August 22, 1823). Cf. G. Abbadessa, 'La filosofia poetica di Vico e Leopardi', *Bollettino del Centro di studi vichiani*, 50 (2020), 219–52.

31 *Zib.*, pp. 164–5 (July 13, 1820).

32 *Zib.*, p. 4129 (April 5–6, 1825)

in *Raymond Sebond's Apology* before Voltaire in the *Contes philosophiques*, Leopardi redefined nature not only as the 'madre benignissima del tutto',[33] a source of illusions that, through the imagination, offers consolation to desire and stimulation to bodily action, but also as a 'persecutrice e nemica mortale', aimed only at the 'distruzione e riproduzione [...] di tutti gl'individui d'ogni genere e specie' rather than at their happiness.[34] As is known, although this vision emerged during and after the writing of the *Operette morali*, already in 1820 Leopardi identified 'un vizio organico [...], radicale, e una contraddizione nel [...] sistema'[35] of nature, noting also that it

> non ci ha dato idea di altri doveri che verso noi stessi ed ha limitato le norme del giusto ai rapporti che l'animale ha con se stesso. Già verso gli animali d'altra specie [...] la natura non ha dettato nessuna regola di onestà e di rettitudine perchè l'uomo non prova nessuna ripugnanza nel far male agli altri animali anche senza suo vantaggio e per mero diletto, come a uccidere una formica ec. E gli altri animali si pascono bene spesso di animali di altra specie. Ma eziandio nella propria specie, l'uomo assolutamente primitivo, non sente infinitamente nessuna colpa a far male a' suoi simili per suo vantaggio, come non la sentono gli altri animali, che maltrattano, combattono, e [...] anche si cibano dei loro simili [...].[36]

Leopardi, however, did not completely abandon his initial philosophical perspective, continuing to recognize nature's role as a 'maestra'.[37] After having applied the theory of pleasure to all living beings as early as 1820, Leopardi distinguished the more sensitive and unhappy ones, as marked by a strong self-love, from the less sensitive ones, to redefine man in 1823 as the least 'sociale di tutti gli altri viventi, [...] perché avendo più vitalità, ha piú amor proprio, e quindi necessariamente ciascun individuo umano ha piú odio verso gli altri individui sí della sua specie sí dell'altre'.[38] In 1823 Leopardi

33 *Zib.*, p. 1530 (August 20, 1821)
34 *Zib.*, pp. 4485–6 (April 11, 1829).
35 *Zib.*, p. 328 (November 14, 1820).
36 *Zib.*, p. 249 (September 19, 1820).
37 *Zib.*, p. 1560 (August 25, 1821).
38 *Zib.*, p. 3773 (October 25–30, 1823).

in fact radicalized the opposition between humans and animals by highlighting the violence that characterizes the former. Before observing, for example, that wolf societies are united in attacking sheep but not in attacking each other,[39] the poet emphasized the dehumanization that characterizes primitive human violence, noting that the ancients animalized their enemies:

> L'inumanità verso i nemici non era biasimo ai tempi di Omero, perchè i nemici non erano considerati come uomini, o come parte di quel corpo a cui apparteneva il loro avversario. Gli antichi (e i selvaggi altresì) erano ben lontani dal considerare tutto il genere umano come una famiglia, e molto piú dal considerare i nemici come loro simili e fratelli. Simili e fratelli non erano per gli antichi, e non sono per li selvaggi, se non gl'individui della loro stessa società; o nazione o cittadinanza o esercito che la vogliamo chiamare [...]. Quindi essere inumano verso i nemici, tanto era per gli antichi, quanto essere *inumano* verso i lupi o altri animali che non sono del genere *umano*, anzi gli nocciono. Siccome appunto i nemici nocevano o cercavano di nuocere a quella società, dentro i limiti della quale si conteneva tutta quella famiglia umana a cui gli antichi si stimavano appartenere. E come a chi prendesse a difendere o a vendicare la sua società contro gli animali nocivi, sarebbe lode il non perdonar loro in alcuna maniera, ma sterminarli tutti a poter suo; cosí agli antichi era lode l'inumanità verso i nemici, che non si reputavano aver diritto all'umanità, non istimandosi aver nulla di umano, cioè nulla di comune con quegli *uomini* che li combattevano; e l'eccesso o il sommo grado di questa inumanità si giudicava proprissima dell'eroe. Massimamente che tutte le passioni o azioni forti erano fra gli antichi stimate molto più degne, o certo più eroiche che le deboli; e quindi la spietatezza verso chi non aveva alcun titolo alla clemenza, quali si stimavano i nemici, era creduta molto più eroica che la compassione, affetto dolce, molle, e stimato femminile; la vendetta molto più eroica che il perdono, siccome il risentimento era giudicato ben più degno dell'uomo che la pazienza delle ingiurie, la quale non andava mai disgiunta dalla riputazione e dal biasimo di viltà o dapocaggine.[40]

Criticizing the culture of archaic Greece, Leopardi alluded *ante litteram* to the tradition of humanism, which in its first meaning

39 *Zib.*, pp. 3781–2 (October 25–30, 1823)
40 *Zib.*, pp. 2765–76 (June 11, 1823)

indicated philanthropy, love for humanity, and which found among its earliest models, before Dante's *Comedy* – in particular the 'umana famiglia' of *Pd.* XXVII, 141, taken up in an anti-humanistic key by Foscolo's *Sepolcri* (v. 5 'bella d'erbe famiglia e d'animali')[41] –, Cicero's studies on the *humana conditio*, also followed by Montaigne.

Adopting a humanist perspective, Leopardi exalted – in harmony with Pico della Mirandola[42] – traditions of thought distant from the so-called Western one, including first and foremost the Jewish one that permeates above all the *Cantico del gallo silvestre*, the text that concluded the first manuscript version of the *Operette morali* (1824), satirically inverting the sacred and biblical language to call readers to awaken from ideologies, in a poetic key that simultaneously overturns, in a materialist and mechanistic perspective, the stoic apocatastasis of Seneca's *Naturales Quaestiones*. It is not necessary to recall that Leopardi's political critique also targeted Christian and conservative culture as responsible for annihilating the needs of the body by projecting happiness into the afterlife; a culture exalted by Italian liberals who, despite suffering foreign domination, continued to celebrate the myth of Optimism and progress, thus evading the issue of human violence, even contemporary violence. At the same time, Leopardi condemned the involvement of Christian culture in colonialism. In number LXVI of his *Pensieri*, composed between 1831 and 1835, the poet observed that:

> Nel secolo presente i neri sono creduti di razza e di origine totalmente diversi da' bianchi, e nondimeno totalmente uguali a questi in quanto è a diritti umani. Nel secolo decimosesto i neri, creduti avere una radice coi bianchi, ed essere una stessa famiglia, fu sostenuto, massimamente da' teologi spagnuoli, che in quanto a diritti, fossero per natura, e per volontà divina, di gran lunga inferiori a noi. E nell'uno e nell'altro secolo i neri furono e sono venduti e comperati, e fatti lavorare in catene

41 G. Abbadessa, 'L'allegoria in Leopardi: l'eco dantesca', *Lexicon Philosophicum. International Journal for the History of Texts and Ideas*, 6 (2018), 133–75 (p. 165).

42 See Giulio Busi, *La Qabbalah* (Rome–Bari: Laterza 2022), pp. 3–4, 64–82.

> sotto la sferza. Tale è l'etica, e tanto le credenze in materia di morale hanno che fare colle azioni.[43]

In his critique of modernity Leopardi renewed the legacy of humanism not only celebrating the primitive poetic language and the ancient philosophical heritage, but also ironically denouncing their oblivion. This is equally evident, for example, in the ode *Ad Angelo Mai* (1820), which pays tribute to Cicero's *De Republica* without ever naming it directly.[44] Even in the draft of the *Dialogo di un lettore di umanità e di Sallustio*, excluded from the *Operette morali*, Leopardi denounced the contemporary world's inability to recognize the virtuous ideals of antiquity, *in primis* that of the homeland, vindicated by Sallust against Catiline.

To condemn the falsehoods propagated in the 'età delle macchine',[45] Leopardi also promoted a nocturnal and demonic humanism, for example with the figure of Momus in the *operetta* entitled *La scommessa di Prometeo* or by depicting Machiavelli as the tutor of Pluto's son in the draft of the *Novella Senofonte e Niccolò Machiavelli*, which satirizes the distortion of language perpetrated by the ruling class. Contrary to what is claimed in 'I libri di morale, di politica, d'eloquenza, di poesia, di romanzi', Machiavelli in fact teaches that '*virtù* significa *ipocrisia*, ovvero *dappocaggine*; *ragione*, *diritto* e simili significano *forza*; *bene*, *felicità* ec. dei sudditi significa *volontà*, *capriccio*, *vantaggio* ec. del sovrano. Cose tanto antiche e note che fa vergogna e noia a ricordarle'[46]. Later, in the first of his *Pensieri*, Leopardi further specified: 'sogliono essere odiatissimi i buoni e i generosi perchè [...] sono sinceri, e chiamano le cose coi

43 G. Leopardi, *Pensieri*, ed. by Antonio Prete (Milan: Feltrinelli 2015), p. 112. For the translation see G. Leopardi, *Thoughts*, trans. by John Gough Nichols (Richmond: Alma Books, 2017).

44 Cf. Costanza Geddes da Filicaia, 'Leopardi, Cicerone, l'"humanitas" e il pensiero linguistico', *Il Capitale Culturale. Studies on the Value of Cultural Heritage*, 24 (2021), 193–204.

45 G. Leopardi, *Operette morali*, ed. by Laura Melosi (Milan: Rizzoli, 2013), p. 141. For the translation see G. Leopardi, *The moral essays*, trans. by Patrick Creagh (New York: Columbia University Press, 1983).

46 Leopardi, *Operette morali*, p. 645.

loro nomi. Colpa non perdonata al genere umano, il quale non odia mai tanto chi fa male, nè il male stesso, quanto chi lo nomina'.[47]

Leopardi never ceased to renew the humanist tradition, as will be further explored in the next paragraph, nor to exalt the model of large animal societies, having also noted in the *Zibaldone* that these gather above all when they perceive a threat: from this perspective—as has been said—partly anarchic and therefore distant from Machiavelli's Renaissance thought, comes the famous invitation of the *Ginestra* (1836) towards the human world to form a 'social catena' against the common enemy of nature.[48] Recognizing in animality a more egalitarian social model than that of humanity, Leopardi thus continued the tradition of alternative humanism that stretched from Montaigne to Erasmus Darwin: it is needless to recall that the poet of Recanati included animals among the protagonists of the *Operette morali*, before the *Paralipomeni*, to criticize the civilization that increases tighter societies in which, as the *Dialogo di Tristano e di un Amico* specifies, '*Gl'individui sono spariti dinanzi alle masse*'.[49]

4. *Between Humanism and Nihilism*

The influence of humanism and, more specifically, of Pico della Mirandola's *Discorso sulla dignità dell'uomo*, also emerges in Leopardi's theory of pleasure. As early as 1820, the poet observed that infinite desire is 'ingenito o congenito coll'esistenza' of all beings but that it provokes in humans 'il sentimento della nullità di tutte le cose'.[50] Meditating on the contrast between the vanity of the infinite desire for pleasure and the human desire for knowledge – recognized in the first book of Aristotle's *Metaphysics* (980a1) – Leopardi, as is known, exemplified this tension in the *Zibaldone* with the fable of Cupid and Psyche and that of the Tree of Genesis. Between 1821 and 1823, following the humanistic tradition, particularly that of

47 Leopardi, *Pensieri*, p. 33.

48 Gilberto Lonardi, *Come i cavalli confederiamoci contro il lupo cattivo*, in *Un mondo 'inconcepibile'*, pp. 37–46.

49 Leopardi, *Operette morali*, p. 597 (Leopardi's italics).

50 *Zib.*, p. 165 (July 12–23, 1820).

Vico, the poet clarified these allegories as spontaneous products of ancient imagination, stimulated by nature, which at that time he explicitly defined as oriented toward a poetic purpose.[51] However, even before this redefinition, Leopardi had already condemned in 1820 the decadence of Humanism in modern civilization, criticizing the denaturation of humanity induced by rationalism:

> E se la natura ha voluto […] la felicità degli esseri, perchè, supponendo che l'abbia posta riguardo all'uomo nella cognizione del vero, ha nascosto questo vero […]? […] Come ha reso così difficile […] la felicità? e la felicità dell'uomo, il quale tiene evidentemente il primo rango nell'ordine delle cose di quaggiù? Come ha ripugnato con ogni sorta di ostacoli a quello ch'ella cercava? Ma l'uomo dovea ben tenere il primo rango, e lo terrebbe anche in quello stato naturale che noi consideriamo come brutale; non però dovea mettersi in un altr'ordine di cose, e considerarsi come appartenente ad un'altra categoria, e porre la sua dignità, non nel primeggiare tra gli esseri, come avrebbe sempre fatto, ma nel collocarsi assolutamente fuori della loro sfera, e regolarsi con leggi apparte, e indipendenti dalle leggi universali della natura.[52]

In 1822, observing that self-love is 'il sentimento universale che abbraccia tutta l'esistenza', but above all the human being, Leopardi went further and described man as the most unhappy of all living beings, thereby profaning the tradition of Renaissance humanism:

> l'uomo avendo per la sua natura […] maggior vita, maggior capacità di più vasta e più numerosa concezione, maggior sentimento insomma, o maggiore sensibilità di tutti gli altri viventi, dee necessariamente avere maggiore intensità, attività, ed estensione o quantità o sentimento d'amor proprio, che non ha verun altro genere di viventi. Quindi l'uomo per essenza propria e inseparabile, è, e nasce più infelice, o meno capace di felicità che verun altro genere di viventi, o di esseri.[53]

51 Cf. *Zib.*, pp. 637–8 (February 10, 1821) and *Zib.*, p. 2939–41 (July 11, 1823).

52 *Zib.*, pp. 327–8 (November 14, 1820).

53 *Zib.*, pp. 2411–12 (May 2, 1822).

In 1825, revisiting some intuitions from 1820,[54] Leopardi shifted his perspective once again, concluding that all sensitive beings 'o la sensibilità astrattamente, sono una imperfezione della natura, o vero gli ultimi, cioè infimi di grado e di nobiltà e dignità nella serie degli esseri e delle proprietà delle cose'.[55] Pico's metaphor of the world as a temple[56] thus becomes a garden of suffering in the *Zibaldone*:

> Non gli uomini solamente, ma il genere umano fu e sarà sempre infelice di necessità. Non il genere umano solamente ma tutti gli animali. Non gli animali soltanto ma tutti gli altri esseri al loro modo. Non gl'individui, ma le specie, i generi, i regni, i globi, i sistemi, i mondi.
>
> Entrate in un giardino di piante, d'erbe, di fiori. Sia pur quanto volete ridente. [...] Tutta quella famiglia di vegetali è in istato di *souffrance*.[57]

Pico's admiration for the condition that human beings have 'in sorte nell'ordine dell'universo [...] secondo le leggi di un arcana sapienza',[58] the work of God, becomes, in Leopardi, admiration for the 'arcano mirabile e spaventoso dell'esistenza universale',[59] that is for mysterious system of nature:

> Ammiriamo dunque quest'ordine, questo universo: io lo ammiro piú degli altri: lo ammiro per la sua pravità e deformità, che a me paiono estreme. Ma per lodarlo, aspettiamo di sapere almeno, con certezza, che egli non sia il pessimo dei possibili.[60]

Nonetheless, as previously noted, Leopardi continued to denounce, even in 1829, the tragedy of the human condition, which must confront the terrible contradiction between 'il fine della natura generale e quello della umana',[61] the former coinciding with 'distruzione e

54 Cf. *Zib.*, pp. 164–5 (July 13, 1820) and note number 31 of this essay.
55 *Zib.*, p. 4134 (April 9, 1825).
56 Pico della Mirandola, *Discorso sulla dignità dell'uomo*, ed. by Francesco Bausi (Parma: Guanda/Pietro Bembo Foundation 2003), p. 7.
57 *Zib.*, p. 4175 (April 22, 1826).
58 Pico della Mirandola, *Discorso sulla dignità dell'uomo*, pp. 5–7.
59 Leopardi, *Operette morali*, p. 472.
60 *Zib.*, p. 4257 (March 24, 1827).
61 *Zib.*, p. 4128 (April 5–6, 1825).

riproduzione',[62] the latter with unattainable happiness. Faced with the modern awareness of man's desperate destiny, as has also been recalled, Leopardi profoundly revisited the tradition of humanism, both by celebrating it in an ironic key and by radically transforming its legacy. On the one hand, the poet overturned the hierarchical system, albeit fluid, recognized by Pico della Mirandola, according to which man occupies a median yet privileged position between angels and beasts. This vision is embodied not only by the *Zibaldone* but also by the *Operette morali*: for example, the solitary philosopher in *Elogio degli Uccelli*, who interprets flight as an animalistic form of human laughter, exalts 'inoperosity', thus outlining a possibility of animal life for man that passes through the suspension of the use of the world.[63] At the same time, this *operetta* satirically denounces, once again, the political crisis of the Restoration, as the philosopher plans to write a

> storia del riso [...] fino a questo tempo presente; nel quale egli si trova essere in dignità e stato maggiore che fosse mai; tenendo nelle nazioni civili un luogo, e facendo un ufficio, coi quali esso supplisce per qualche modo alle parti esercitate in altri tempi dalla virtù, dalla giustizia, dall'onore e simili; e in molte cose raffrenando e spaventando gli uomini dalle male opere.[64]

Instead, the *Il Copernico. Dialogo* satirically evokes the scientific revolution of modernity: in this *operetta*, as is known, the allegorical figure of the Sun charges Copernicus to reveal to humanity that he no longer intends to revolve around the earth, and the eponymous protagonist protests, declaring that such a change 'sconvolgerà i gradi della dignità delle cose, e l'ordine degli enti, scambierà i fini con le creature; e per tanto farà un grandissimo rivolgimento anche nella metafisica, anzi in tutto quello che tocca alla parte speculativa del sapere'.[65] Faced with the collapse and scientific overcoming of

62 *Zib.*, p. 4486 (April 11, 1829).
63 Alessandra Aloisi, 'Elogio dell'inoperosità: Agamben e Leopardi', *Italian Studies*, 22.10 (2017), pp. 1–11.
64 Leopardi, *Operette morali*, p. 450–1.
65 Ibid., p. 532.

Renaissance culture embodied in Pico della Mirandola's *Discourse*, the Sun, however, responds: 'diversamente da Cicerone, ho riguardo più all'ozio che alla dignità'.[66] This statement recalls what Leopardi observed in *Parini ovvero della gloria*: recognizing that 'il soggetto principale delle lettere è la vita umana', this *operetta* alludes with 'dignity' to the figure of Cicero and to the classical ideal of *otium* which, as is well known, was conceived in ancient Rome as a time removed from the public time of *negotium*, but equally aimed at 'glory' understood, in Leopardi's own words, as 'beneficio della repubblica'.[67] While recalling the humanist tradition through the case of the famous Roman lawyer and rhetorician, the *operetta* however recognized that 'i moderni [...] che si pongono per via degli studi [...] non hanno necessità di scusarsi colla loro patria'.[68] Leopardi thus denounced the political and cultural apathy of Italy, showing how the few who commit themselves to the pursuit of glory, and therefore of the benefit of the country, were in reality neglected and misunderstood by a public that was no longer able to read, or rather, to compete with the printing industry, founded on quantity rather than quality.

In the *Pensieri*, and in particular in number LXVIII, Leopardi's humanism reached a peak, integrating into Renaissance thought the desperate philosophy implied by his theory of pleasure:

> La noia è in qualche modo il più sublime dei sentimenti umani. [...] Il non poter essere soddisfatto da alcuna cosa terrena, né, per dir così, dalla terra intera; considerare l'ampiezza inestimabile dello spazio, il numero e la mole maravigliosa dei mondi, e trovare che tutto è poco e piccino alla capacità dell'animo proprio; immaginarsi il numero dei mondi infinito, e l'universo infinito, e sentire che l'animo e il desiderio nostro sarebbe ancora più grande che sì fatto universo; e sempre accusare le cose d'insufficienza e di nullità, e patire mancamento e voto, e però noia, pare a me il maggior segno di grandezza e di nobiltà, che si vegga della natura umana. Perciò la noia è poco nota agli uomini di nessun momento, e pochissimo o nulla agli altri animali.[69]

66 Ibid.
67 Ibid., pp. 298–9.
68 Ibid., p. 301.
69 Leopardi, *Pensieri*, p. 114.

After criticizing 'dignity' as an obsolete concept, Leopardi used its synonym 'nobility'[70] to redefine the modern humanism through boredom. In the *Elogio degli Uccelli* boredom was defined as the desire for pleasure left in its pure state, and for Alessandra Aloisi, following Giorgio Agamben, it indicated a uniquely human dimension that reclaims its animality.[71] Conversely, in the sceptical and nihilistic humanism of the *Pensieri*, boredom also indicates a possible dimension shared by humans and animals, although Leopardi emphasizes the rarity of this experience—especially among animals. Clearly, the poet embraced the polysemy of poetic language, far from the lexicon employed by Heidegger and later Agamben to distinguish 'noia profonda' from 'stordimento animale'.[72] The detailed description in *Thought* LXVIII, however, suggests that Leopardi's noble boredom bears no resemblance to animal captivation; rather, it approximates the deep boredom analyzed by Heidegger, albeit from a wholly human perspective that Agamben only very partially contested.[73] Althought Aloisi limits her analysis to Leopardi's *Elogio degli Uccelli* and his poetry, following Agamben, she incisively points out that, for the poet of Recanati, 'nella noia [...], il desiderio è sì tenuto in sospeso, ma la sua potenza non è stata ancora convertita a un nuovo possibile uso, perché allora il desiderio non sarebbe più noia, ma riso, canto, danza, contemplazione'.[74] Leopardi's noble boredom is, in fact, a feeling of nothingness that, less widespread and less human than laughter, similarly deactivates the use of the world, though without

70 Leopardi uses the two lemmas in a synonymous way, for example, in *Zib.*, p. 4493 (April 22, 1829).

71 Aloisi, 'Elogio dell'inoperosità', p. 3, cf. Giorgio Agamben, *L'uso dei corpi* (Vicenza: Neri Pozzi, 2014), p. 240.

72 G. Agamben, *L'aperto. L'uomo e l'animale* (Turin: Bollati Boringhieri, 2002), pp. 70–1. For the translation by Kevin Attel cf. G. Agamben, *The Open: Man and Animal* (Stanford: Stanford University Press, 2002).

73 G. Agamben (*L'aperto*, pp. 73–4) observes that even the 'animal può effettivamente – in circostanze particolari come quelle cui l'uomo lo sottopone nei laboratori – sospendere la relazione immediata col suo ambiente, senza per questo cessare di essere un animale né diventare umano'.

74 Aloisi, 'Elogio dell'inoperosità', p. 10.

converting it into a new possible use – that is, without transforming the landscape into a site of pure potential. Unlike laughter, noble boredom derives from the imagination of a beyond-landscape: an imagination that embraces and suspends all natural space to reveal, akin to laughter, the state of inoperosity of the living being.[75]

75 Rather than being 'una prova dell'incapacità umana di eguagliare l'inoperosità divina' (Ibid., p. 10, n. 53), noble boredom reveals that 'dal nulla scaturiscono le cose che sono' (Leopardi, *Operette morali*, p. 468).

Roberta Visone*

MAN AS AN EXCEPTION TO NATURE IN THE REFLECTIONS OF ALFRED RUSSEL WALLACE

Abstract

This paper deals with the subject of man in the thought of the Welsh naturalist Alfred Russel Wallace (1823–1913), who, together with Charles Darwin, discovered the theory of the origin of species through natural selection. By analysing some of Wallace's works, I trace the stages of his gradual departure from Darwinian theory, which finds its greatest expression in the elaboration of a doctrine of the 'exceptionality' of man in comparison with other living beings.

Keywords

Evolution – Wallace – Natural selection – Anthropogenesis – Design argument

At the Special General Meeting of the Linnean Society of London on 1 July 1858, two papers signed by Charles Darwin (1809–1882) and the young Welsh naturalist Alfred Russel Wallace (1823–1913) were presented, officially introducing the doctrine of the modification of species by natural selection to the scientific community.[1] Darwin and Wallace had independently

* Università degli Studi di Napoli Federico II.

1 Charles Darwin and Alfred Russel Wallace, 'On the Tendency of Species to Form Varieties; and on the Perpetuation of Varieties and Species by Natural Means of Selection. Communicated by Sir Charles Lyell and Joseph D. Hooker to the LSL Meeting of 1 July 1858', *Journal of the Proceedings of the Linnean Society of London. Zoology*, 3 (1858), 53–62.

discovered the mechanism that would revolutionise the concept of the biological universe, and finally put the creationist paradigm in check, and to which Darwin dedicated his 1859 masterpiece *The Origin of Species*.[2]

Darwin's epistemological break with traditional explanations of biological phenomena consisted above all in the introduction of a revolutionary theoretical element, 'chance', which undermined the traditional approach with which naturalists had interpreted the organic universe until then: natural selection, which causes species to change over time, is an anti-teleological principle; it does not operate according to predetermined or predictable plans, but on the basis of organic variations that arise in organisms. The emergence of new species is presented as the product of contingent factors, as the 'accidental' coincidence of two independent causal chains:[3] the one that leads to an organism exhibiting a certain variation, and the other that comprises the set of circumstances that characterise the environment outside the organism, i.e. the scenario of its struggle for survival. New species emerge through the differential survival of individuals that at a given moment are endowed with an accidental variation that gives them an advantage in this struggle, that they pass on to their offspring.

The fact that the variations that determine the survival of a given organism are accidental does not mean that the evolutionary process is left to pure chance, to the simple insignificance of infinite random fluctuations: there are laws that govern the process, but – and this is the point that interests Darwin – they do not make the process

---

2 Charles Darwin, *On the origin of species by means of natural selection, or the preservation of favoured races in the struggle for life* (London: Murray, 1859).

3 Telmo Pievani describes the Darwinian evolutionary mechanism as the fortuitous encounter of two independent causal chains, the first being the source of internal contingency (individual variations) and the second being the source of external contingency (the external conditions of existence between species and environment and between species and species): cf. Telmo Pievani, *Anatomia di una rivoluzione: la scoperta scientifica di Darwin* (Milan–Udine: Mimesis, 2020), p. 71.

predictable, nor do they direct it along necessary lines so that it could conceivably be the project of a deliberate mind.[4]

The partly accidental nature of the evolutionary processes described by Darwin was one of the main reasons why many scientists, even those who called themselves 'Darwinists', were reluctant to accept the doctrine of natural selection. In *One Long Argument* (1991), Ernst Mayr sets out a scheme in which he identifies the basic theoretical elements of Darwinism that were incorporated into some of the major evolutionary theories (descent from a common ancestor, gradualism of the evolutionary process, population speciation, natural selection) between the second half of the nineteenth and the beginning of the twentieth century, showing that no scientist other than Darwin fully accepted the argument from natural selection.[5] Even Herbert Spencer, who gained worldwide fame as the 'philosopher of evolution' thanks to his *System of Synthetic Philosophy* published between the 1860s and 1890s,[6] ruled out natural selection as the only driving force behind the evolution of species. He denied the 'spontaneous' nature of the variations on which it acts and relegated it to a secondary factor in evolution, which he believed could be explained primarily by Lamarck's principles of use/disuse and the inheritance of acquired characteristics.

Since his early work of 1850, *Social Statics*, in which naturalistic and ethical-political considerations are interwoven with the classical themes of natural theology, according to which nature is the product of a 'divine idea', Spencer presents 'progress' (for him synonymous with 'evolution') as a universal law whose goal is

4 In other words, the actual result of the organic forms is not a necessary one, but only one of many possible results (cf. ibid., p. 71).

5 Mayr's scheme includes the theories of Lamarck, Darwin, Haeckel, the neo-Lamarckians, T. Huxley, H. De Vries, and T. H. Morgan. See Ernst Mayr, *One Long Argument: Charles Darwin and the Genesis of Modern Evolutionary Thought* (Cambridge, MA: Harvard University Press, 1991), p. 37.

6 *A System of Synthetic Philosophy* consists of ten volumes divided into: *First Principles, Principles of Biology, Principles of Psychology, Principles of Sociology, Principles of Ethics.* See Robert G. Perrin, *Herbert Spencer: A Primary and Secondary Bibliography* (London and New York: Routledge, 1993), pp. 116–25.

cosmic equilibrium or the ever more perfect adaptation of organisms to their environment. Progress, according to Spencer, is achieved through the transition from a primitive state of indeterminate and incoherent homogeneity to a state of determinate and coherent heterogeneity, leading to an ever more perfect balance between the 'inside' of the organism (its structures and functions) and its 'outside' (the environment), and this applies to lichens as well as to animals and societies.

Spencer's notion of progress has a necessary, teleological and ameliorative character: 'Progress [...] is not an accident, but a necessity', it says in *Social Statics*,[7] and this characterisation can be traced back to the embryological genesis of the spencerian idea of progress. Spencer's 'law' of progress is in fact inspired by the 'law of increasing heterogeneity' discovered by the Estonian zoologist Karl von Baer in the 1830s, according to which the development of animal embryos is characterised by a functional and structural 'increasing heterogeneity' (the organs of the embryo, initially unspecialised, gradually differentiate to establish a perfect balance between the internal organisation of the organism and the environment).[8] In *Social Statics*, which reflects Spencer's anti-statist and liberal ideological positions, the law of progress is extended to the ethical and political sphere by establishing a 'first principle' of justice in accordance with the progressive structure of the universe, from which all individual rights, especially the right to freedom, are derived: this principle is the law of 'equal freedom'. Only by following the rules of this paradigm of 'justice', which prescribes

7 Herbert Spencer, *Social statics, or, The Conditions Essential to Human Happiness Specified, and the First of Them Developed* (London: Chapman [1850], 1851), p. 65.

8 Karl Ernst von Baer's theory is set out in his work *Über Entwickelungsgeschichte der Thiere. Beobachtung und Reflexion*, 2 vols (Königsberg: Bornträger, 1828–1837), which was partially translated for the British public by Thomas Huxley in the early 1850s under the title 'On Development of Animals, with Observations and Reflections', in *Scientific Memoirs: Selected from the Transactions of Foreign Academies of Sciences, and from Foreign Journals. Natural History*, ed. by Thomas Henry Huxley and Arthur Henfrey (London: Taylor and Francis, 1853), pp. 176–238.

the definitive recognition of the centrality of the social atom, the individual, as the subject of inalienable rights and freedoms, is it possible to follow the path of social progress.

In the *System*, published a few years after *The Origin of Species*, Spencer abdicates his role as a political philosopher and the natural theology approach of *Social Statics* in order to adhere to the scientific codes of the debate on the modification of species triggered in Europe by the 'Darwinian revolution'. In *Social statics*, however, the 'law of evolution', the driving force behind the change of species at the organic level, retains its finalist-embryological roots and comes into open conflict with the Darwinian interpretation of life. Spencer attributes the causes of biological evolution not so much to the occurrence of accidental variations, but rather to the ability of the individual to exercise its faculty to adapt to the environment by obeying the dictates of the law of evolution, which enables it to 'win' the battle for existence. This conviction also stems from Spencer's youthful interest in a discipline that was already considered a pseudoscience in the mid-19th century: phrenology. According to the latter, which was concerned with the 'necessary laws' of mental processes, the exercise of the 'faculties' with which an organism is endowed inevitably leads to an ever more perfect adaptation of the organism to its environment.

In the *Principles of Biology* (volumes II and III of the *System*, published between 1864 and 1867), Spencer therefore recognises that Darwin's mechanism of natural selection is of secondary importance compared to Lamarck's principles of use/disuse and inheritance of acquired characteristics, which are more in line with his phrenological beliefs: according to these principles, it could be claimed that evolution takes place thanks to the ability of the biological individual to adapt to its environment through the use of certain faculties that are fixed in species by inheritance. In brief, the 'winners' in the 'struggle for existence' are the individuals and groups that are equipped with a stronger faculty of adaptation.[9]

9 See Franco Ferrarotti, *Lineamenti di storia del pensiero sociologico* (Rome: Donzelli, 2002), p. 97.

Spencer's works, especially *Social Statics,* were almost a cult object for Wallace.[10] In his 1864 essay *The Origin of Human Races*, the co-author of the theory of natural selection stated, in exquisitely Spencerian terms, that 'the power of "natural selection"' had led 'to the more perfect adaptation of man's higher faculties to the conditions of surrounding nature, and to the exigencies of the social state'.[11] In this essay, Wallace describes man as an 'exception' to nature due to his intellectual and moral dispositions, which put an end to the changes in his physical structure caused by natural selection.[12] With these statements, to which he says he was led by 'the perusal of Mr. Herbert Spencer's works, especially *Social Statics*',[13] Wallace partially changed the content of the doctrine of natural selection, which he had presented to the Linnean Society just six years earlier. The mind now represents a clear separation that distinguishes man from the rest of nature, it is a part of the 'new series of causes'[14] that intervene in natural systems and interrupt the effects of natural selection. Unlike other animals – Wallace explains – which are more vulnerable or unable to survive in the face of changing environmental conditions, humans have been able to survive thanks to their mental power, using, for example, clothing and weapons to compensate for climate change and biotic competition, or fire and agriculture to free themselves from fluctuations in food availability, while the

10 In his autobiography Wallace recognises the significant influence that *Social statics* had on his thinking. See Alfred Russel Wallace, *My life: A record of events and Opinions*, 2 vols (London: Chapman and Hall, 1905), II, p. 235, and ibid., p. 267. To get an idea of the extent of Wallace's admiration for Spencer, it is worth remembering that the Welsh naturalist named his eldest son, who was born in 1867, 'Herbert Spencer'.

11 Alfred Russel Wallace, 'The Origin of Human Races and the Antiquity of Man Deduced from the Theory of "Natural Selection"', *Journal of the Anthropological Society of London*, 2 (1864), clviii–clxxxvii (hereafter Wallace 1864), p. clxix. See Spencer, *Social statics,* p. 59, p. 85, p. 153, p. 315.

12 See Malcolm Jay Kottler, 'Alfred Russel Wallace, the Origin of Man, and Spiritualism', *Isis*, 65.2 (1974), 144–92 (p. 147).

13 Wallace 1864, p. clxx.

14 Ibid., p. clxiv.

emergence of moral sentiments has gradually prevented the weakest from being at the mercy of the harsh law of natural selection.

Wallace's final break with Darwin on natural selection took place in 1869–1870, and this divergence was also influenced by the Welsh naturalist's new experiences in the field of psychical research.[15] In 1869 Wallace published an essay on human evolution in the *Quarterly Review.*[16] In it, he explained that he had changed his mind from his earlier essay of 1864: not only did he now claim, as he had in that essay, that man's intellectual powers had emancipated him from the action of natural selection, but he also hypothesized that the 'great laws of organic development in the origin of the human race as in the origin of all organized beings' seems 'to be evidence of a Power which has guided the action of those laws in definite directions and for special ends'.[17]

This hypothesis was put forward because it was difficult to explain by the theory of natural selection alone certain characteristics that have been typical of our species since its appearance on the Earth and for which prehistoric man could have had no real need,[18] such as the size of the brain and the vocalization of the larynx, among others that seemed to be downright disadvantageous, such as the lack of body hair.[19]

In a letter, Wallace had given Darwin a preview of the topics he would cover in his 1869 article: 'In my forthcoming article in the "Quarterly"', he wrote, 'I venture for the *first time* on some limitations to the power of natural selection [which] are the expression of a

15 See Giacomo Scarpelli, *Il cranio di cristallo. Evoluzione della specie e spiritualismo* (Turin: Bollati Boringhieri, 1993), p. 11 and Kottler, 'Alfred Russel Wallace', p. 188. From the mid-1960s, Wallace regularly took part in spiritualist and mesmeric practises, and these experiences may also have contributed to the genesis his 'evolutionary teleology'. Martin Fichman, 'Science in Theistic Contexts: A Case Study of Alfred Russel Wallace', *Osiris*, 16 (2001), 227–50 (p. 228).

16 A. R. Wallace, 'Sir Charles Lyell on geological climates and the origin of species', *The Quarterly Review*, 126 (1869), 359–94 (hereafter Wallace 1869).

17 Wallace 1869, p. 393.

18 See ibid., p. 391.

19 See ibid., pp. 392–3.

deep conviction founded on evidence which I have not alluded to in the article but which is to me absolutely unassailable'.[20] Darwin replied that he was 'intensely curious' to read the article, but added resignedly: 'I hope you have not murdered too completely your own & my child'.[21] Darwin's premonition could not have been more accurate: in order to explain the emergence of consciousness and moral sense in the human species, i.e. the appearance of the human mind,[22] Wallace stated in his article that 'the mind of the man' cannot be regarded as the result of natural selection, but must rather be explained as being 'itself the living proof' of the existence of a 'Higher Intelligence'.[23]

By pointing to a higher intelligence as the cause of the human plane, Wallace, as S. J. Gould has noted, did not deny the central role of natural selection in the evolutionary process, but rather affirmed it with greater vigour: it was precisely his 'peculiarly rigid view of natural selection' as the sole cause of evolutionary change that 'led him to reject it for the human mind'.[24] Wallace therefore places everything that cannot be explained by natural selection outside the natural level, since the latter would only determine a character in relation to its immediate utility. To give an example, in Wallace's view the peculiarities of the human brain, whose size in individuals from primitive societies far exceeds their actual needs, cannot be regarded as a product of natural selection, since it has not undergone such changes in volume over time that it could be

---

20 *The correspondence of Charles Darwin*, ed. by Frederick H. Burkhardt et al. (Cambridge: Cambridge University Press, 1985–), hereafter CCD, vol. XVII, p. 155 (Alfred Russel Wallace's letter to Charles Darwin, March 24, 1869).

21 Ibid., p. 157 (Charles Darwin's letter to Alfred Russel Wallace, March 27, 1869).

22 Here Wallace discusses the specificity of man in comparison to other living beings – namely man's 'mental specificity', which 'today would be called cultural' specificity (see Scarpelli, *Il cranio di cristallo*, p. 45) – which he had already discussed in his essay of 1864, in which he did not, however, explicitly attribute it to a divine origin.

23 Wallace 1869, p. 394.

24 See Stephen Jay Gould, *The Panda's Thumb. More Reflections in Natural History* (New York–London: Norton, 1992), pp. 53–4.

argued that there is a significant difference between savages from primitive societies and civilised man.[25] For Wallace, this meant that the human brain cannot be a product of selection (which 'could only have endowed the savage with a brain a little superior to that of an ape, whereas he actually possesses one but very little inferior to that of the average members of our learned societies'),[26] but is due to a superior intelligence that has predetermined a special destiny for man, making him suitable for needs and requirements that would arise only in the future.[27] Similarly, the presence of complex speech organs in the primitive man is interpreted as 'an instrument' that 'has been developed in advance of the needs of its possessor'.[28]

The hypothesis put forward by Wallace in 1869 that organic processes are controlled by a higher intelligence suggests the idea that there is a supernatural 'selector' – an idea that the naturalist had decisively ruled out just a few years earlier. Due to his anti-finalist view of natural processes, in 1866, Wallace had even viewed the parallelism between natural and artificial selection developed in *The Origin of Species* with suspicion and suggested that Darwin himself replaced the term 'natural selection' in later editions of his work with the term 'survival of the fittest' introduced by Spencer in his *Principles of Biology*. In Wallace's view the personification of nature suggested by Darwin's equation of natural selection with artificial selection could indeed have exposed the doctrine of natural selection to some misunderstandings such as teleological interpretations and ultimately evoked the image of a selecting being – an 'intelligent chooser'[29] –, whereas Spencer's expression seemed better suited to describe natural processes because of its more impersonal and objective character.[30]

The revisions that Wallace made to the theory of natural selection in 1869 – which, in his opinion, proved capable of reconciling

25 See Wallace 1869, p. 391.
26 Ibid., p. 392.
27 See Wallace 1869, pp. 391–2.
28 Ibid., p. 393.
29 CCD, vol. XIV, p. 228 (Wallace's letter to Darwin, July 2, 1866).
30 See ibid., pp. 227–8.

science and theology in relation to the 'most momentous problem',[31] i. e. anthropogenesis – enabled him to recall the parallelism between natural and artificial selection as a paradigm for the possible action of a higher intelligence that, over time, like a patient breeder, has guided the laws of the evolution of organic forms for a 'nobler end', the generation of the human species together with its mental and moral dispositions:

> Man himself guides and modifies nature for special ends. The laws of evolution alone would perhaps never have produced a grain so well adapted to his uses as wheat; such fruits as the seedless banana, and the bread-fruit; such animals as the Guernsey milch-cow, or the London dray-horse. [...] We must [...] admit the possibility, that in the development of the human race, a Higher Intelligence has guided the same laws for nobler ends. Such, we believe, is the direction in which we shall find the true reconciliation of Science with Theology on this most momentous problem. [...] Let us not shut our eyes to the evidence that an Overruling Intelligence has watched over the action of those laws, so directing variations and so determining their accumulation, as finally to produce an organization sufficiently perfect to admit of, and even to aid in, the indefinite advancement of our mental and moral nature.[32]

Darwin openly rejected these claims, as the marginal notes he wrote on Wallace's article in his copy of the *Quarterly Review* show.[33] After reading it, he immediately communicated his impressions to

---

31 Wallace 1869, p. 394.

32 Ibid., pp. 393–4.

33 Several notes that Darwin wrote on his copy of the *Quarterly* show his dissatisfaction with his colleague's reasoning, e.g. a thrice underlined 'No!' in the margin of Wallace's passage on the mental needs of savages, or four exclamation marks next to the passage in which Wallace argues the impossibility to ascribe the development of the brain in savages to natural selection alone). See, with regard to this point: Joel Schwartz, 'Darwin, Wallace, and the "Descent of man"', *Journal of the History of Biology*, 17.2 (1984), 271–89 (p. 277); Ross Slotten, *The Heretic in Darwin's Court: The Life of Alfred Russel Wallace* (New York: Columbia University Press, 2004), pp. 269–70; Michael Flannery, 'Alfred Russel Wallace's Intelligent Evolution and Natural Theology', *Religions*, 11, 316 (2020), 1–19 (p. 8).

the author: 'I differ grievously from you, and I am very sorry for it. I can see no necessity for calling in an additional and proximate cause in regard to Man'.[34]

Wallace replied that his opinions on the subject 'have been modified' from what they had been in the past in virtue of 'the consideration of a series of remarkable phenomena, physical & mental', which he now had 'every opportunity of fully testing, and which demonstrate the existence of forces and influences not yet recognised by science'.[35]

Wallace took up the question of anthropogenesis again in two essays in the collection *Contributions to the Theory of Natural Selection* (1870),[36] which he published also in anticipation of the forthcoming publication of Darwin's *The Descent of Man* and in which he continued to argue in favour of the exceptionality of man in comparison with the rest of nature. In these writings, some of which deals with the subjects of the essays of 1864 and 1869,[37] the existence of the moral sense – one of the main themes of Spencer's *Social Statics* – becomes proof of the inadequacy of the explanation of natural selection in relation to the origin of man. While the practise of moral feelings such as honesty can still be traced back to

34 CCD, vol. XVII, p. 175 (Charles Darwin's letter to Alfred Russel Wallace, April 14, 1869).

35 Wallace adds that he is aware that these statements could appear to be the result of a 'mental hallucination', see James Marchant, *Alfred Russel Wallace. Letters and Reminiscences*, 2 vols (London–New York–Toronto–Melbourne: Cassel, 1916) I, p. 244 (Alfred Russel Wallace's letter to Charles Darwin, April 18, 1869). Wallace had already alluded to these phenomena in his previous letter of March 24, describing them as the 'absolutely unassailable' evidence on which his 'deep conviction' about the origin of man is rooted (see note 20 above).

36 Alfred Russel Wallace, *Contributions to the Theory of Natural Selection. A Series of Essays by Alfred Russel Wallace* (London: Macmillan, 1870).

37 The collection contains a reprint with some additions of the 1864 essay on the development of human races (*The Development of Human Races under the Law of Natural Selection*, see *Contributions*, pp. 302–31) and an essay that takes up the subject of the exceptionality of man due to his intellectual and moral nature, which Wallace had addressed in his 1869 essay (*The Limits of Natural Selection as Applied to Man*, cf. ibid., pp. 332–71).

natural selection, the ‘higher feelings of pure morality’, the ‘power of abstract reasoning and ideal conception’[38] which qualify man as a ‘spiritual being’[39] cannot be explained in utilitarian terms, nor can the innate ‘sense of right and wrong’ and the ‘acts of universal utility’ or ‘self-sacrifice’: these expressions of moral sense cannot be the result of a selection process because of their lacking of a direct biological utility.[40]

In *The Descent of Man* (1871) Darwin admitted that one of the two objects he had in view in *The Origin of Species,* in addition to revealing the mechanism of natural selection, had been ‘to overthrow the dogma of separate creations’.[41] The rejection of this dogma, however, he remarked, must be accompanied by the abandonment of a strictly utilitarian account of natural selection – the tacit assumption that ‘every detail of structure’ of an organism is ‘of some special, though unrecognised, service’[42] – since this account still derives from a conception of the organic world as a product of an intentional creation. In his book on man, Darwin reflected on the importance of recognising the role of natural selection within a non-teleological conception of the history of the organic world, and when he described those who, on the basis of teleological assumptions, tend to ‘extend the action of natural selection, either during past or present times, too far’,[43] he seemed to be referring to Wallace. However, he was not referring to the co-discoverer of the theory of natural selection, but to a ‘metamorphosed’ Wallace, as he described him in a letter he wrote after reading his colleague’s anthropological essays. Wallace’s ‘metamorphosis’ did not bode well, as it seemed to be – as Darwin jokingly remarked – going ‘in retrograde direction’.[44]

---

38 Wallace, *Contributions*, p. 356.

39 Ibid., p. 359.

40 See ibid.

41 Charles Darwin, *The Descent of Man*, 2 vols (London: John Murray, 1871), I, p. 153.

42 Ibid.

43 Ibid.

44 CCD, XVIII, p. 17. Darwin ends his letter of January 26, 1870 to Wallace signing himself ‘Your miserable friend’.

Starting from a hyper-adaptationist perspective, Wallace came to the seemingly paradoxical conclusion that all phenomena that cannot be explained by the adaptive effects of natural selection should be regarded as supernatural. If, in the case of *Homo sapiens*, selection alone was not able to explain the acceleration of brain development found by palaeontology, if selection alone was not able to explain the superiority of human mental abilities, then in Wallace's eyes man appeared as a being necessarily separate from others: 'he is [...] a being apart, since he is not influenced by the great laws which irresistibly modify all other organic beings'.[45]

In 1870, the Swiss zoologist René-Édouard Claparède sharply criticised the hypothesis of the 'higher intelligence' with which Wallace, in his opinion, had completely modified his previous theory of natural selection, turning man into a 'God's domestic animal'.[46] On his side, Darwin was more lenient towards the tendency he found in the writings of many of his contemporaries to deny the 'lower origin' of man in order to place him above other living beings. Although it does not lead to the path of truth, as we read in the epilogue to *The Descent of Man*, this tendency seems to be rooted in human nature itself and can therefore only appear to us as a forgivable mistake.

> Man may be excused for feeling some pride at having risen, though not through his own exertions, to the very summit of the organic scale; and the fact of his having thus risen, instead of having been aboriginally placed there, may give him hopes for a still higher destiny in the distant future. But we are not here concerned with hopes or fears, only with the truth as far as our reason allows us to discover it. I have given the evidence to the best of my ability; and we must acknowledge, as it seems to me, that man with all his noble qualities, with sympathy which

45 Wallace 1864, p. clxviii (see also Wallace 1870, p. 326).

46 Wallace, *Contributions to the Theory of Natural Selection*, 2nd edition (London: Macmillan, 1871), p. 372. Wallace uses this expression in the second edition of the *Contributions,* in response to the negative criticism of his work by Claparède, who had remarked that the Welsh naturalist in his anthropological writings seemed to claim that 'les races humains constitueraient [...] les animaux domestiques d'une force divine'. René-Édouard Claparède, 'La sélection naturelle et l'origine de l'homme', *Revue des cours scientifiques*, 7, 36 (1870), 564–71 (p. 569).

> feels for the most debased, with benevolence which extends not only to other men but to the humblest living creature, with his god-like intellect which has penetrated into the movements and constitution of the solar system – with all these exalted powers – Man still bears in his bodily frame the indelible stamp of his lowly origin.[47]

In the course of his life, which he devoted largely to field research,[48] Wallace published hundreds of scientific treatises on zoology, biology, botany, geology, climatology and biogeography. The methodical rigour with which he observed nature was never at odds with his conviction, developed in the 1860s, that the natural world could be the product of a supernatural design. In his book *Island Life* (1880), paraphrasing the poet Alexander Pope, he stated he was firmly convinced that an ever more thorough exploration of natural phenomena would only strengthen our confidence 'that the "mighty maze" of Being we see everywhere around us is "not without a plan"'.[49]

---

47 Darwin, *The Descent*, II, p. 405.

48 From 1848 to 1852 Wallace lived in the Amazon region, and from 1854 to 1862 he settled in Southeast Asia to continue his exploration trips.

49 Alfred Russel Wallace, *Island Life or, the Phenomena and Causes of Insular Faunas and Floras* (London: Macmillan, 1880), p. 512. In *An Essay on Man* Alexander Pope looks at nature as 'a mighty maze but not without a plan' (Alexander Pope, *An Essay on Man: Being the First Book of Ethic Epistles. To Henry St. John, L. Bolingbroke* (London: Gilliver, 1734), p. 9 [I, 6]).

Raffaele Carbone*

PROUST IN DARWIN'S GARDEN
Understanding Humans through Nature

Abstract

The article investigates Darwin's presence in *In Search of Lost Time* with a view to shedding light on the extent to which the interest in the theories elaborated by the English naturalist and, in general, the vision of the relations between man and nature that emerged after the publication of his works in late Nineteenth-century European culture, had an impact on the way Proust analysed and attempted to account for certain aspects and attitudes of human beings. After reconstructing the context of Darwinian reception in France and of the training of the young Marcel Proust, the article focuses on the first pages of *Sodom and Gomorrah*, in which we find the most elaborate reference to Darwin in Proust's entire oeuvre, in relation to the homosexual encounter between the characters Monsieur de Charlus and Jupien; the article then develops a number of reflections on the presence of classical-literary references and scientific theories in these same pages and in other passages in the monumental seven-volume novel.

Keywords

Darwinian world – Fertilisation of the flowers – Homosexuality – Myth – Science

* Università degli Studi di Napoli Federico II.

1. *Introduction*

1.1 *Proust and the nineteenth century*

We would first like to point out that, according to certain scholars such as Antoine Compagnon, *À la recherche du temps perdu* is an anachronistic novel.[1] It was published between 1913 and 1927, but, apart from the parts added during the First World War, most of its conceptual and scientific references belong to the late 19th century, in particular, to the time when Proust started writing in student journals. According to Céline Surprenant, this element must be taken into account when investigating the possible channels of encounter between the scientific and literary work of Darwin and Proust.[2] Therefore, even though *Du côté de chez Swann* was published in 1913 and the other volumes after the war, the novel was marked by the 'fin de siècle' and contained a 'decadent bric-à-brac', so much so that when the later volumes appeared after the war, the work was received as 'a monument from another era ['un monument d'une autre ère]' or 'a prehistoric monster rooted in the roaring twenties" ['un monstre préhistorique, échoué dans les années folles]'.[3] In addition to the temporal gap between the date of publication and its scientific and cultural content, there is another interesting element to note in relation to the background of the Proustian cycle: a discrepancy between the narrator's fictional theories and the narration. The *Recherche* departs from the French naturalist texts of the late 19th century, in which theories were put at the service of the moral or political teachings of the works, yet there is no lack of theories and laws in it. The narrator seeks 'general laws' for everything and sees in the ability to formulate them, a sign of understanding reality.

1 Antoine Compagnon, *Proust entre deux siècles* (Paris: Seuil, 1989), p. 56.

2 Céline Surprenant, 'Darwin and Proust', in *The Literary and Cultural Reception of Charles Darwin in Europe*, ed. by Thomas F. Glick and Elinor Shaffer (London: Bloomsbury, 2014), pp. 431–57 (p. 434).

3 Compagnon, *Proust entre deux siècles*, pp. 32 and 26, respectively. However, it is not only because of its genesis that the Proustian cycle has been called 'entre deux siecles', but also because of its structures, themes and style. Ibid., pp. 49–50.

This search takes us back to the positivist perspective of the late 19th century, in which human behaviour was thought to obey a set of 'laws'.[4] However, the novel creates a probabilistic universe in which there are multiple motivations and explanations.[5]

The abundance of 'laws' in the novel is accompanied by the positivist element of Proustian literary aesthetics, an aspect that has sometimes been obscured because critics have tended to emphasise the narrator's idealism.[6] Anne Henry, for example, regarded the novel as the faithful narrative transposition of an idealistic worldview, the sources of which are to be found in the philosophies of Schopenhauer and Schelling.[7] The issue, however, seems more complex. Jean-Pierre Ollivier's book, *Proust et les sciences*, is useful in this regard. The author explains that Proust certainly grew up in an intellectual context in which he lived and breathed positivist air: the education of his father – who was a neurologist before obtaining the chair of hygiene at the Faculty of Medicine in Paris – was marked by the positivism of Comte (1798–1857) and the scientism of Marcelin Berthelot (1827–1907).[8] In the unfinished novel *Jean Santeuil* (written between 1895 and 1900), when speaking of M. Santeuil, Jean's father (who represents Proust's own father), the narrator emphasises the 'unreasoning and haughty positivism [le positivisme irraisoné et superbe]' that had been one of his life's illusions.[9]

---

4 Donald Wright, *Du discours médical dans 'A la recherche du temps perdu': Science et souffrance* (Paris: Champion, 2007), p. 224.

5 Compagnon, *Proust entre deux siècles*, p. 50.

6 Jean-Marc Quaranta, 'Génétique et intertextualité', in *Proust au tournant des siècles*, I, ed. by Bernard Brun and Juliette Hassine, series 'Marcel Proust', 4, 'La Revue des Lettres Modernes' (Paris–Caen: Lettres modernes Minard, 2004), pp. 53–74 (p. 56). See also Surprenant, 'Darwin and Proust', p. 437.

7 Anne Henry, *Marcel Proust: Théories pour une esthétique* (Paris: Klincksieck, 1981).

8 Jean-Pierre Ollivier, *Proust et les sciences* (Paris: Champion, 2018), p. 18.

9 M. Proust, *Jean Santeuil* précédé de *Les Plaisirs et les jours*, ed. by Pierre Clarac with the collaboration of Yves Sandre (Paris: Gallimard, 1971), p. 865. Here and hereafter, if the official English translation is not indicated, the translation is our own.

In the preface of his book *Origines de l'alchimie* (1885), which ranks among his works on the history of science, Marcelin Berthelot, who held the chair of Organic Chemistry at the School of Pharmacy in 1859 and at the Collège de France in 1864, affirms:

> The world today is devoid of mystery: the rational conception claims to illuminate and understand everything: it strives to give a positive and logical explanation of all things and extends its fatal determinism to the moral world. I do not know whether the imperative deductions of scientific reason will one day realise this divine prescience, which has given rise to so much discussion in the past and which has never been reconciled with the no less imperative feeling of human freedom. In any case, the entire material universe is claimed by science, and no one dares to resist this claim any longer. The notion of the miraculous and the supernatural has vanished like a vain mirage, an outdated prejudice.[10]

This kind of attitude and approach played a certain role in Proust's early training. His father, Adrien Proust (1834–1903), professeur agrégé of medicine, was an internationally recognised expert on hygiene and infectious diseases. His works include a *Traité d'hygiène*, which was published in 1877 and saw two further editions while the author was alive. In the preface to the first edition, which was also reprinted in later editions, Adrien Proust emphasises a broad meaning of 'hygiene,' which includes everything that can contribute to man's physical and moral well-being[11] and then

10 Marcelin Berthelot, *Origines de l'alchimie* (Paris: Georges Steinheil, 1885), pp. v–vi.

11 'Il est une autre manière de comprendre l'hygiène, plus haute et plus large. Avec elle le programme de cette science s'étend singulièrement; il ne s'agit plus d'un but purement préventif et prophylactique, d'un rôle surtout défensif: tout ce qui peut conduire à l'amélioration de l'homme, à l'accroissement de son bien-être physique et moral, de son activité somatique et intellectuelle, devient du ressort direct et légitime de l'hygiène. Ainsi envisagée, elle franchit les limites étroites de la médecine; et la biologie, l'anthropologie, la législation, l'histoire entière de l'humanité, se réunissent pour constituer le fonds et comme le domaine propre de cette science. Tout ce qui touche à l'homme appartient à l'hygiéniste; il n'a le droit de se désintéresser de rien, et il peut s'appliquer

develops a series of reflections on humanity, evolution and races. It is interesting to note that in the history of human evolution, particularly of western civilisation, Adrien Proust emphasises three epochal moments: (1) the 'great revolution of the Renaissance [grande révolution de la Renaissance]' which opposes the 'sterile mysticism of the Middle Ages [stérile mysticisme du moyen âge]' and was able 'to restore the importance and dignity of real life and earthly things [restituer à la vie réelle et aux choses terrestres leur importance et leur dignité]'; (2) the 18th century, which, compared to the 'abstract and serene heights [hauteurs abstraites et sereines]' of 17th-century metaphysics, produced a philosophy that was 'more restless and militant; it inquires into everything, is interested in and tackles everything that concerns man [plus inquiète et plus militante]; elle s'enquiert de tout, s'intéresse et s'attaque à tout ce qui touche à l'homme]';[12] (3) the 19th century, marked by unprecedented scientific and technological progress, thanks to which it would be possible to combat – according to Adrien Proust – famines and epidemics in every corner of the world.[13]

While the influence of familial and cultural positivism had left a deep mark on the young Proust, which would not disappear in his adult writing, in 1889, at the conclusion of his philosophical education 'en classe de philosophie', thanks to the influence of Professor Alphonse Darlu at the Lycée Condorcet[14] and French philosophy, permeated with spiritualism and idealism, his adherence to positivism and scientism was mitigated.[15]

---

la pensée du poète: *Nihil humani a me alienum puto*'. Adrien Proust, *Traité d'hygiène*, revised and augmented third edition (Paris: Masson et Cie, 1903), 'Préface de la première édition', pp. 5–6).

12 Ibid., p. 8.

13 Ibid, p. 9.

14 On Darlu and Proust, see Luc Fraisse, *L'éclectisme philosophique de Marcel Proust* (Paris: PUPS, 2013), pp. 59, 131–5.

15 Ibid., p. 19.

1.2 *Proust, the sciences and Darwin's reception in France*

One issue that should be explored further when addressing the problem of Darwin's presence in Proust's work concerns the relationship between Proust and the sciences. Fundamental in this respect is the already mentioned work by Ollivier, *Proust et les sciences*, which emphasises Proust's interest in the scientific knowledge of his time both through his family and schooling and through his social life.[16] We do not know exactly what the degree and tenor of Proust's scientific knowledge was in his younger years, but we can identify precise references to scientific theories or laws in the *Recherche*. Consider, among possible examples, the metaphorical use of the first law of thermodynamics, an important discovery of 19th century physics, in *Albertine disparue*.[17]

As regards Proust's familiarity with evolutionary theory, it is essential to take stock of Darwin's reception in France.

In this regard, Gillian Beer wrote that at the end of the 19th century 'everyone found themselves living in a Darwinian world',[18] and this implied that references to evolutionary theory circulated widely, not to mention the explicit assimilation of Darwinian ideas by writers such as Paul Bourget (1852–1935) and Alphonse Daudet (1840–1897), naturalist writers such as Émile Zola (1840–1902) or sociologist Gabriel Tarde (1843–1904). Motivating elements of his fiction with the reference to the 'survival of the fittest', Bourget, for instance, reduced evolutionary theory to the idea that 'the weaker members of

16 Ollivier, *Proust et les sciences*, p. 20.

17 'Comme le mal que j'avais fait à ma grand-mère, le mal que m'avait fait Albertine fut un dernier lien entre elle et moi et qui survécut même au souvenir, car, avec la conservation d'énergie que possède tout ce qui est physique, la souffrance n'a même pas besoin des leçons de la mémoire; ainsi un homme qui a oublié les belles nuits passées au clair de lune dans le bois, souffre encore des rhumatismes qu'il a pris'. Marcel Proust, *À la recherche du temps perdu*, ed. by Jean Yves Tadié, 4 vols (Paris: Gallimard, 1987–1989), vol. 4, p. 107 (henceforth *Recherche*).

18 Gillian Beer, *Darwin's Plots: Evolutionary Narrative in Darwin, George Eliot and Nineteenth-Century Fiction* (London: Routledge & Kegan Paul, 1983), p. 3.

society should be left to their fate'.[19] The debates in literary criticism on the influence of Darwinism on morality involved, among the most important authors, Ferdinand Brunetière (1849–1906) and Anatole France (1844–1924), who debated Bourget's novel *Le Disciple* and Alphonse Daudet's melodrama *La Lutte pour la vie* (both published in 1889).[20]

Marcel Proust's assimilation of the 'Darwinian world' differs from that of these writers. The echoes and few explicit references to Darwin in Proust do not derive from the Spencerian inflection of the theory of natural selection that dominated the French reception of Darwin at the end of the 19th century and that Bourget's positions represented. *À la recherche du temps perdu* contains no elements of reflection on the impact of the theory of evolution on morality. Certainly, it is a sign – over a period of time from the end of the 19th century to the First World War – of the presence of Darwin in the middle and educated classes of French and European society, as a popularised scientific reference that had now entered everyday language.[21]

2. *Proust's knowledge of Darwin's work*

Scholars of Darwin's reception in literary texts or of the relationship between Proust and the sciences, have traced the different ways in which Darwin's work had an impact on the *Recherche*. Proust came into contact with Darwinian ideas both through his readings of great English (George Eliot's *Middlemarch*) and French authors (Daudet's *Lutte pour la vie*) and through the ideas that circulated in his family and cultural milieu: consider in particular the figure of Dr Samuel Pozzi, the French translator of *The Expression of Emotions in Man and Animals*, who was close to Proust's family.[22]

---

19 Louise Lyle, 'Le Struggleforlife: Contesting Balzac through Darwin in Zola, Bourget and Barres', *Nineteenth-Century French Studies*, 36.3–4 (2008), 305–19 (p. 305).

20 Surprenant, 'Darwin and Proust', p. 431.

21 Ibid., p. 432.

22 Ollivier, *Proust et les sciences*, p. 190.

Proust's knowledge of Darwin's work included, with regard to evolutionary theory, *On the Origin of Species by Means of Natural Selection, or the Preservation of Favoured Races in the Struggle for Life* (1859); concerning the plant world and fertilisation processes, *On the Various Contrivances by which British and Foreign Orchids Are Fertilised by Insects*, and *On the Good Effects of Intercrossing* (1862), translated into French by Louis Rérolle in 1870 (Paris, C. Reinwald) and *The Different Forms of Flowers on Plants of the Same Species* (1877), translated into French by Édouard Heckel under the title *Des différentes formes de fleurs dans les plantes de la même espèce* (1878), with an analytical preface by Amédée Coutance (1824–1895). Proust drew upon this preface to write the parts in the *Recherche* in which he used the fertilisation model in the plant world to introduce the ways that homosexual relationships are established.[23]

Consider in this regard a manuscript version of the beginning of *Sodome and Gomorrhe I* (which is different from the text published in May 1921):

> Perhaps before dinner, when Mme de Guermantes returned and I went to see her, she would agree to take me into her garden. Ever since she had interested me in the fertilisation of flowers, I had read Darwin's books [...].[24]

This is a famous passage, extracted precisely from the first pages of *Sodome and Gomorrhe I*: even in the version intended for publication, it is the most explicit and articulate theoretical reference to Darwin in the novel. The narrator illustrates the ways in which the homosexual encounter between the characters Charlus and Jupien is prepared and takes shape by drawing on Darwin's botanical work on the self-fertilisation of orchids. It is significant that here

23 See Emily Eells, 'Darwin', in *Dictionnaire de Marcel Proust*, ed. by Annick Bouillaguet and Brian G. Rogers, new revised edition (Paris: Champion, 2014), pp. 285–6.

24 'Peut-être avant le diner quand Mme de Guermantes serait rentrée et que j'irai la voir consentirait-elle à me mener dans son jardin. Depuis qu'elle m'avait intéressé à la fécondation des fleurs j'avais lu les livres de Darwin [...]'. *Recherche*, III, *Sodome et Gomorrhe*, I, Notes et variantes, p. 1267.

Proust avoided the ideological weight of evolutionary theory while dealing with a subject as complex and scabrous as homosexuality, or 'inversion', as he puts it.[25]

Another significant sphere in which to trace Darwin's presence in the *Recherche* concerns the bodily expressions of the emotions in movement that Proust abundantly portrayed in his cycle of novels and that Darwin studied in *The Expression of the Emotions in Man and Animals*, first published in London in 1872 and translated into French in 1874 by René Benoit (1844–1921) and Samuel Pozzi (1846–1918), who worked in the medical circle of Proust's father. It was re-edited in 1877 and 1890.[26] Even without empirical evidence that Proust read this work by Darwin, there are remarkable points of intersection between the two authors around the theme of the expression of emotions, such as the Lamarckian role of habit and the relationship between emotions and the past, to mention only these points, involving physiology and aesthetics.[27]

3. *The reference to Darwin in* Sodome et Gomorrhe I

The most elaborate reference to Darwin in *À la recherche du temps perdu* is to be found in the opening thirty pages of *Sodome et Gomorrhe*, published in two parts in 1921 and 1922, which was the result of the expansion of the initial project of a three-volume novel during the interruption caused by the war. *Sodome et Gomorrhe* 1 and 2 deal with the theme of homosexuality. The story narrated in *Sodome et Gomorrhe I* constitutes a kind of initiation equally important as the madeleine at the beginning of *Du côté de chez Swann*: from here on, the theme of homosexuality would emerge prominently, and male homosexuality in particular would be perceived in every place

25 Consider also the explicit evocation of Darwin's name in *Le Temps retrouvé* with regard to how flowers are fertilised. *Recherche*, IV, p. 360.

26 See Surprenant, 'Darwin and Proust', p. 433.

27 Janet Browne, *Darwin and the Expression of the Emotions*, in *The Darwinian Heritage*, ed. by David Kohn (Princeton: Princeton University Press, 1985), pp. 307–26 (p. 310).

and social context, in Paris as in Balbec, by the Guermantes as by the Verdurins.[28]

In 1909, Proust had sketched out pieces on the theme of 'sexual inversion', entitled 'La race maudite' (The accursed race) and borrowed the term 'La race des tantes' from Balzac, to refer to 'homosexuals',[29] involving M. de Guercy, who would later become M. de Charlus.

In the first pages of *Sodome et Gomorrhe*, the narrator tells of an episode preceding his visit to the duke and duchess of Guermantes (already narrated in the previous book). Before the visit, he was already in their palace and, moving from the belvedere at the top of the building up the stairs, was spying on their return. Involuntarily, while his attention had been caught by a shrub and the orchid in the courtyard (which he could observe from the louver on the staircase), the narrator made an important discovery that he considers essential to illustrate at this point in the narrative.[30] This occurred while he was wondering whether 'the unlikely insect would come, by a providential hazard, to visit the offered and neglected pistil [l'insecte improbable viendrait, par hazard providentiel, visiter le pistil offert et délaissé]',[31] this insect that Amédée Coutance, in the 'Préface' to *Des différentes formes de fleurs dans les plantes de la même espèce*, defines as a 'trigger of opportunities [générateur d'occasion]'.[32]

The encounter, a Jupien-flower awaiting the arrival of a Charlus-bumble-bee, described by drawing on a botanical model inspired by Darwin, 'distracts' the narrator from the fertilisation of the flower by the insect, but in reality the botanical references in these pages are the means by which he describes an encounter between human

28 See the 'Notice' of *Sodome et Gomorrhe*, in *Recherche*, III, p. 1186.

29 *Recherche*, III, p. 919.

30 *Recherche*, III, p. 1.

31 *Recherche*, III, p. 1; *In Search of Lost Time* (henceforth *SLT*), vol. 4: *Sodom and Gomorrah*, ed. by William C. Carter, trans. by C. K. Scott Moncrieff (New Haven and London: Yale University Press, 2021), p. 4 (henceforth *SaG*).

32 Amédée Coutance, 'Préface', in Charles Darwin, *Des différentes formes de fleurs dans les plantes de la même espèce*, French translation by Édouard Haeckel (Paris: C. Reinwald, 1878), pp. ix–xxxvi (p. xxvii).

beings. Human sexuality is thus introduced and somewhat veiled by the sexuality of flowers. Botanical knowledge becomes the condition for perceiving the meaning of the 'ritual preliminaries [préludes rituels]'[33] that the narrator witnesses and for deciphering what is happening between the two men.

In the comparison between Charlus and Jupien's encounter and the fertilisation of the orchid by the bumble-bee, already suggested in *Esquisse II*,[34] but developed in the final version, Proust drew inspiration from Maeterlinck's *L'intelligence des fleurs* (Paris: Fasquelle, 1907) and Darwin's works: *De la fécondation des orchidées par les insectes et des bons résultats du croisement* (trans. by L. Rérolle, Paris: C. Reinwald, 1870); *Des effets de la fécondation croisée et de la fécondation directe dans le règne végétal* (trans. by Édouard Haeckel, C. Reinwald, 1877); *Des différentes formes de fleurs dans les plantes de la même espèce* (trans. by Édouard Haeckel, C. Reinwald, 1878). In fact, all Proust's information seems to derive from the aforementioned 'Preface' by Prof. Amédée Coutance to *Des différentes formes de fleurs dans les plantes de la même espèce*.[35]

Let us proceed by recalling the narrative context in which the references to Darwin are inscribed. Baron de Charlus is one of the main characters in the *Recherche*: an exemplary widower who worships his deceased wife like no other; he is a homosexual who hides his desires from others and who cross-dresses.[36] The narrator

33 *Recherche*, III, p. 8; *SaG*, p. 8.

34 *Recherche*, III, p. 938.

35 Coutance, 'Préface', pp. ix–xxxvi.

36 See *La Prisonnière*: 'Il ne voulait pas voir que, depuis dix-neuf cents ans ("un courtisan dévot sous un prince dévot eût été athée sous un prince athée", a dit La Bruyère), toute l'homosexualité de coutume – celle des jeunes gens de Platon comme des bergers de Virgile – a disparu, que seule surnage et se multiplie l'involontaire, la nerveuse, celle qu'on cache aux autres et qu'on travestit à soi-même. Et M. de Charlus aurait eu tort de ne pas renier franchement la généalogie païenne. En échange d'un peu de beauté plastique, que de supériorité morale! Le berger de Théocrite qui soupire pour un jeune garçon, plus tard n'aura aucune raison d'être moins dur de cœur, et d'esprit plus fin, que l'autre berger dont la flûte résonne pour Amaryllis. Car le premier n'est pas atteint d'un mal, il obéit aux modes du temps. C'est l'homosexualité survivante malgré les obstacles,

presents him as an 'error of nature'[37] struggling in vain against a tendency represented as an irreversible genetic determinism.[38] Having positioned himself at the top of the inner staircase of the building of the Duchess of Guermantes, the narrator sets out to observe through the shutters what is going on in the garden, where there are plants, insects and human beings (M. de Charlus visiting his aunt, Mme de Villeparisis and the 'ancien giletier', the farmer Jupien). At first, the narrator awaits the arrival of an insect that makes the fertilisation of the flower possible by bringing in the seed of another flower (which is far more effective than the self-fertilisation of the flower, which actually causes degeneration and sterility). Before dwelling on this aspect of flower fertilisation, Proust wrote: 'The laws of the vegetable kingdom are themselves governed by increasingly higher laws [Les lois du monde végétal sont gouvernées elles-mêmes par des lois de plus en plus hautes]'.[39] However, the narrator is unable to see the miracle of fertilisation take place because his attention is drawn to the encounter between M. de Charlus and Jupien: the former, who did not imagine that he would find Jupien, looked at him 'with unusual attention [avec une attention extraordinaire]', while the latter, 'suddenly rooted to the spot, like a plant, before M. de Charlus, was contemplating with a look of amazement the plump form of the aging baron [cloué subitement sur place devant M. de Charlus, enraciné comme une plante, contemplait d'un air émerveillé l'embonpoint du baron vieillissant]'.[40] M. de Charlus's demeanour having changed, Jupien's demeanour harmonized with it, as if obeying the laws of an occult art.[41] A few lines later, the narrator notes:

honteuse, flétrie, qui est la seule vraie, la seule à laquelle puisse correspondre chez le même être un affinement des qualités morales' (*Recherche*, III, p. 710).

37 *Recherche*, III, p. 300.

38 See Edward J. Hughes, 'Charlus (baron de)', *Dictionnaire de Marcel Proust*, pp. 204–6.

39 *Recherche*, III, p. 5; *SaG*, p. 5.

40 *Recherche*, III, p. 6; *SaG*, pp. 6–7 (translation amended).

41 *Recherche*, III, p. 6; *SaG*, p. 7.

> Meanwhile Jupien, shedding at once the humble, virtuous air that I had always associated with him, had – in perfect symmetry with the baron – thrown back his head, given a becoming tilt to his body, placed his hand with a grotesque impertinence on his hip, stuck out his behind, struck poses himself with the coquetry that the orchid might have adopted on the providential arrival of the bumblebee.[42]

And still later, the narrator thus describes Mr de Charlus and Jupien, who seemed to have concluded their agreement:

> Nearer still to nature – and the multiplicity of these analogies is itself all the more natural in that the same man, if we examine him for a few minutes, appears in turn as a man, a man-bird or man-insect, and so forth – one would have called them a pair of birds, the male and the female, the male seeking to advance, the female – Jupien – no longer giving any sign of response to these overtures, but regarding her new friend without surprise, with an inattentive fixity of gaze, which she doubtless felt to be more disturbing and the only effective method, once the male had taken the first steps, and contented herself with preening her feathers. [...] At the same instant, just as M. de Charlus disappeared through the gate humming like a great bumblebee, another, a real one this time, came into the courtyard. For all I knew this might be the one so long awaited by the orchid, and that was coming to bring it that rare pollen without which it must remain a virgin. But I was distracted from following the frolics of the insect for, a few minutes later, engaging my attention anew, Jupien (perhaps to pick up a parcel, which he did take away with him later, and which, presumably, in the emotion aroused by the apparition of M. de Charlus, he had forgotten, or perhaps simply for a more natural reason) returned, followed by the baron. The latter, deciding to cut short the preliminaries, asked the tailor for a light, but at once observed: 'I ask you for a light, but I see I've forgotten my cigars'. The laws of hospitality prevailed over those of coquetry. 'Come inside,

42 'Or Jupien, perdant aussitôt l'air humble et bon que je lui avais toujours connu, avait – en symétrie parfaite avec le baron – redressé la tête, donnait à sa taille un port avantageux, posait avec une impertinence grotesque son poing sur la hanche, faisait saillir son derrière, prenait des poses avec la coquetterie qu'aurait pu avoir l'orchidée pour le bourdon providentiellement survenu'. Ibid.

you will be given everything you desire', said the tailor, on whose features disdain now gave place to joy.[43]

Another significant passage in *Sodome et Gomhorre I* is found towards the end of the section. It is a general reflection on 'inversion', a term Proust uses to designate homosexuality, preferring it to the latter. 'Homosexual' is a term coined in 1869 by the Hungarian physician Karoly Maria Benkert, known as Kertbeny, in two anonymous pamphlets[44] and later used in German psychiatry.[45] However, Proust preferred the term used by Charcot, who designated

43 'Plus près de la nature encore – et la multiplicité de ces comparaisons est elle-même d'autant plus naturelle qu'un même homme, si on l'examine pendant quelques minutes, semble successivement un homme, un homme-oiseau ou un homme-insecte, etc. – on eût dit deux oiseaux, le mâle et la femelle, le mâle cherchant à s'avancer, la femelle – Jupien – ne répondant plus par aucun signe à ce manège, mais regardant son nouvel ami sans étonnement, avec une fixité inattentive, jugée sans doute plus troublante et seule utile, du moment que le mâle avait fait les premiers pas, et se contentant de lisser ses plumes. […] Au même instant où M. de Charlus avait passé la porte en sifflant comme un gros bourdon, un autre, un vrai celui-là, entrait dans la cour. Qui sait si ce n'était pas celui attendu depuis si longtemps par l'orchidée, et qui venait lui apporter le pollen si rare sans lequel elle resterait vierge? Mais je fus distrait de suivre les ébats de l'insecte, car au bout de quelques minutes, sollicitant davantage mon attention, Jupien (peut-être afin de prendre un paquet qu'il emporta plus tard et que dans l'émotion que lui avait causée l'apparition de M. de Charlus, il avait oublié, peut-être tout simplement pour une raison plus naturelle), Jupien revint, suivi par le baron. Celui-ci, décidé à brusquer les choses, demanda du feu au giletier, mais observa aussitôt: "Je vous demande du feu, mais je vois que j'ai oublié mes cigares". Les lois de l'hospitalité l'emportèrent sur les règles de la coquetterie. "Entrez, on vous donnera tout ce que vous voudrez", dit le giletier, sur la figure de qui le dédain fit place à la joie. La porte de la boutique se referma sur eux et je ne pus plus rien entendre'. *Recherche*, III, pp. 8–9; *SaG*, pp. 8–9.

44 See Wayne R. Dynes (ed.), *Encyclopedia of Homosexuality*, 2 vols (Chicago/London: St James Press, 1990), vol. 1, p. 555; Jean Claude Féray and Manfred Herzer, 'Kertbeny, une énigmatique "mosaïque d'incongruités"', *Études finno-ougriennes*, 22 (1990), 215–39.

45 See Richard Freiherr von Krafft-Ebing, *Psychopathia sexualis. Eine klinisch-forensische Studie* (Stuttgart: Ferdinand Enke, 1886).

homosexuality as 'inversion of the sexual instinct'.[46] In the late 19th and early 20th century, European societies were marked by a number of scandals related to male homosexuality, the best known being that of Oscar Wilde, who was sentenced to hard labour in 1895. In the following year, some fundamental essays were published: Marc-André Raffalovich, *Uranisme et unisexualité: Étude sur différentes manifestations de l'instinct sexuel* (Paris: Masson, 1896); Georges Saint-Paul, *Tares et poisons: Perversion et perversités sexuelles*, preface by Émile Zola (Paris: Carré, 1896); Havelock Ellis, *Sexual Inversion* (Watford/London: University Press, 1897). In *L'Eau sous les ponts* Paul Morand states that Proust was familiar with the work of Magnus Hirschfeld, who, following in the footsteps of Ulrichs, proposed the theme of the third sex.[47] As Pauline Moret-Jankus states, '[i]t is undeniable, in fact, that Proust adopts the theory of the man-woman, according to which the male homosexual is in reality a woman: a woman imprisoned in a male body'.[48]

In an addition to the *Cahier 49*,[49] the narrator sets out his linguistic, but not exclusively linguistic, reasons for the choice of the term 'inverted': the inverted (also preferred to 'tante', a somewhat trivial word) is a man who loves men because he is, in fact, a woman. The homosexual (i.e., according to these definitions, a man who is a man and loves men) does not exist.

> Finally, inasmuch as inversion itself springs from the fact that the invert is too closely akin to woman to be capable of having any effective relations with her, it comes under a higher law that ordains that so many hermaphrodite flowers will remain unfertile, that is to say the law of the sterility of self-fertilization. It is true that inverts, in their search for a male, often content themselves with other inverts as effeminate as themselves. But it suffices that they do not belong to the female sex, of which they have in them an embryo that they can put to no useful purpose, as happens in so many hermaphrodite flowers, and even in certain hermaphrodite animals, such as the snail, which cannot

---

46 *Recherche*, III, *Sodome et Gomorrhe*, I, Notes et variantes, p. 1278.

47 Paul Morand, *L'Eau sous les ponts* (Paris: Grasset, 1954), p. 54.

48 Pauline Moret-Jankus, *Race et imaginaire biologique chez Proust* (Paris: Classiques Garnier, 2016), p. 193.

49 *Recherche*, III, *Sodome et Gomorrhe*, I, *Esquisse* IV, p. 955.

be fertilized by themselves, but can be by other hermaphrodites. In this respect the race of inverts, who eagerly connect themselves with Oriental antiquity or the golden age in Greece, might be traced back farther still, to those experimental epochs in which there existed neither dioecious plants nor monosexual animals, to that initial hermaphroditism of which certain rudiments of male organs in the anatomy of women and of female organs in that of men, seem still to preserve the trace. I found the pantomime, incomprehensible to me at first, of Jupien and M. de Charlus as curious as those seductive gestures addressed, Darwin tells us, to insects by the flowers called composite that erect the florets of their capitula so as to be seen from a greater distance, such as a certain heterostyled flower that turns back its stamens and bends them to open the way for the insect, or which offers him an ablution, and, even quite simply, the nectar fragrance and vivid hue of the corollas that were at that moment attracting insects to our courtyard.[50]

---

50 'Enfin, l'inversion elle-même venant de ce que l'inverti se rapproche trop de la femme pour pouvoir avoir des rapports utiles avec elle, se rattache par là à une loi plus haute qui fait que tant de fleurs hermaphrodites restent infécondes, c'est-à-dire à la stérilité de l'autofécondation. Il est vrai que les invertis à la recherche d'un mâle se contentent souvent d'un inverti aussi efféminé qu'eux. Mais il suffit qu'ils n'appartiennent pas au sexe féminin, dont ils ont en eux un embryon dont ils ne peuvent se servir, ce qui arrive à tant de fleurs hermaphrodites et même à certains animaux hermaphrodites, comme l'escargot, qui ne peuvent être fécondés par eux-mêmes, mais peuvent l'être par d'autres hermaphrodites. Par là les invertis, qui se rattachent volontiers à l'antique Orient ou à l'âge d'or de la Grèce, remonteraient plus haut encore, à ces époques d'essai où n'existaient ni les fleurs dioïques ni les animaux unisexués, à cet hermaphroditisme initial dont quelques rudiments d'organes mâles dans l'anatomie de la femme et d'organes femelles dans l'anatomie de l'homme semblent conserver la trace. Je trouvais la mimique, d'abord incompréhensible pour moi, de Jupien et de M. de Charlus aussi curieuse que ces gestes tentateurs adressés aux insectes, selon Darwin, par les fleurs dites composées, haussant les demi-fleurons de leurs capitules pour être vues de plus loin, comme certaine hétérostylée qui retourne ses étamines et les courbe pour frayer le chemin aux insectes, ou qui leur offre une ablution, et tout simplement même que les parfums de nectar, l'éclat des corolles, qui attiraient en ce moment des insectes dans la cour'. *Recherche*, III, pp. 30–1; *SaG*, pp. 34–5.

Two interesting references are discernible in this passage, one more hidden, the other obvious due to the clear evocation of Darwin: (1) the first concerns the myth recounted by Aristophanes in the *Symposium* (the three types of men of the origins: one endowed with two male bodies, one endowed with two female bodies, and the androgyne); (2) the other reference is to Darwin's conception that the division of the sexes occurred late in the evolution of species, which raised the question of whether this theory allowed Proust to justify the inversion with the idea of an early hermaphroditism. It should also be considered that in *Albertine disparue* Proust presents the homosexuality of another character, Legrandin, as 'a kind of return to nature, albeit in a roundabout way [une sorte de retour, même détourné, vers la nature]'.[51]

The theory of evolution is effective on at least two levels. On the one hand, it contributes to the 'inversion' theory elaborated by the narrator and based on the analogy between the sexuality of plants and that of humans. On the other, it inspires the discovery of aesthetic truths.

Let us consider this passage:

> This Romeo and this Juliet may believe with good reason that their love is not the caprice of a moment but a true predestination, prepared by the harmonies of their temperaments, and not only by their own personal temperaments but by those of their ancestors, by their most distant strains of heredity, so much so that the person who is conjoined with them has belonged to them from before their birth, has attracted them by a force comparable to the one that governs the worlds on which we spent our former lives.[52]

---

51 *Recherche*, IV, p. 246.

52 'Ce Roméo et cette Juliette peuvent bien croire à bon droit que leur amour n'est pas le caprice d'un instant, mais véritable prédestination préparée par les harmonies de leur tempérament, non pas seulement par leur tempérament propre, mais par celui de leurs ascendants, par leur plus lointaine hérédité, si bien que l'être qui se conjoint à eux leur appartient avant la naissance, les a attirés par une force comparable à celle qui dirige les mondes où nous avons passé nos vies antérieures'. *Recherche*, III, p. 29; *SaG*, pp. 32–3.

It is worth emphasising here the reference to the biological inheritance in which previous lives endure.

The 'miraculous' encounter[53] had been announced at the end of *Le Côté de Guermantes* as the discovery of 'a landscape no longer Turneresque but moral ['un paysage non plus turnérien mais moral]'.[54] However, with the encounter encoded in *Sodome et Gomorrhe I*, the dimension of aesthetics emerges: the beauty of the encounter becomes central.[55] Darwin allows Proust to combine the viewpoints of natural history and aesthetics, as did the naturalist himself.[56]

The theory that the narrator proposes explains the composition of the sexes in light of 'initial hermaphroditism [hermaphroditisme initial]', traces of which are found in 'certain rudiments of male organs in the anatomy of women and of female organs in that of men [quelques rudiments d'organes males dans l'anatomie de la femme et d'organes femelles dans l'anatomie de l'homme]'.[57] The dual composition of the sexes leads one to believe that the choice of sexual partner is only apparently based on whether or not one possesses a male or female character. Darwin's botanical work on orchids is evoked insofar as it is the law that governs 'the sterility of self-fertilization [la stérilité de l'auto-fécondation]',[58] and that also governs the conjunction witnessed by the narrator. Indeed, it is not so much the union between two men that is sterile, but rather that between an inverted man and a woman. The latter can be sterile because the inverted man comes too close to the woman, harbouring a female embryo.[59]

53 'Mais c'était un miracle aussi auquel je venais d'assister, presque du même genre, et non moins merveilleux'. Ibid.

54 *Recherche*, II, p. 861.

55 *Recherche*, III, p. 29.

56 Jonathan Smith, *Charles Darwin and Victorian Visual Culture* (Cambridge: Cambridge University Press, 2006), pp. 166–7.

57 *Recherche*, III, p. 31; *SaG*, p. 35.

58 *Recherche*, III, p. 31; *SaG*, p. 34. On this point, see Simon Porzak, 'Inverts and Invertebrates: Darwin, Proust and Nature's Queer Heterosexuality', *Diacritics*, 41.4 (2013), 6–34.

59 *Recherche*, III, p. 31; *SaG*, p. 34. On this point, see Surprenant, 'Darwin and Proust', p. 441.

4. *Conclusion: Myth and science*

Science, and the new image of nature and man that it had drawn in the second half of the 19th century, constitute a privileged channel through which Proust represents and seeks to understand the attitudes and characteristics of human beings. From time to time, the classical tradition also surfaces.

In conclusion, we would like to return to the incipit of *Sodome et Gomhorre I*, specifically to a different version of the beginning, found in the manuscript notebook and in the copy of the typed text. In this version, already mentioned, there is an explicit reference to the reading of Darwin's books.

> Perhaps before dinner, when Mme de Guermantes returned and I went to see her, she would agree to take me into her garden. Since she had interested me in the fertilisation of flowers I had read Darwin's books [...].[60]

It is also interesting to note a reference to classical mythology.

> Pour tout dire par cette chaude journée, ce que j'étais altéré de voir plus que des fleurs, plus que le ciel avec ses nuées, c'était ces autres nuages colorés qui presque à ras du sol voyagent de fleurs <en fleurs>; pareilles à ces gerbes prismatiques d'eau que lancent les tuyaux d'arrosage pour rafraîchir les jardins; peut-être dans celui de Mme de Guermantes pourrais-je voir de vraies fusées de pollen projetées par les fleurs anémophiles; la pluie de soufre qu'égouttent, abondante, fine et serrée, les fleurs du pin, les nuages d'or qui voyagent entre les genêts d'Espagne, tout ce monde aussi inconnu de moi, presque aussi invisible pour mon œil qui n'avait jamais su le remarquer que les ondes électriques qui nous entourent, et dont j'aurais tant voulu qu'un Elstir, de son regard subtil et révélateur qui avait bien su percer le crépuscule et capter les vapeurs, eût fixé pour m'apprendre à les reconnaître, champ encore inexploré de la peinture, les brumes errantes et colorées, les averses intermittentes, attendues, instantanées et voulues comme un geste décisif, providentielles et fécondes comme la pluie d'or de Sémélé (?).[61]

60 *Recherche*, III, *Sodome et Gomorrhe*, I, Notes et variantes, p. 1267.

61 Ibid., p. 1268.

In the variant, the narrator expresses, in a more baroque and poetic manner, his desire to observe the phenomenon of flower fertilization. In particular, he expresses his desire to see a flare of pollen launched by anemophilous flowers. The reference here is to anemophilous pollination, a type of abiotic pollination that uses the wind as a means of dispersal. Proust could always find this reference in Darwin's work translated into French by Édouard Haeckel: *Des différentes formes de fleurs dans les plantes de la même espèce*.[62] The narrator is attracted and fascinated by the phenomenon of flower fertilization: a world unknown to him that could have been represented pictorially by the artist Elstir, a world interwoven with gaseous and liquid elements that emanate and wander and which the narrator compares to the golden rain of Semele (seduced by Zeus and mother of Dionysus), although the reference is to Danae, mother of Perseus, with whom Zeus conjoined after introducing himself in the form of a golden rain into one of his father's towers. The rain of gold is the culmination of a complex system of metaphors that starts from the pollen flares launched by the flower through the wind and other atmospheric phenomena.

It is equally significant that, in the concluding version, the allusion to the Semele/Danae myth disappears, as does the direct reference to Darwin, but the naturalistic-Darwinian background plot remains as the key to the world of homosexual encounters. In the variant of the incipit, scientific references and references to classical mythology coexist on the same page, while in the print version the mythical-classical reference disappears. Who knows if unconsciously the narrator wishes to tell us that, in order to understand human beings, it is necessary to use no longer, or not only, the *studia humanitatis* but, above all, the study of nature, updated by the most recent discoveries and investigations.

Nevertheless, as we have seen, myth is not absent from these pages. It emerges there where Proust deals with the theme of hermaphroditism, the coexistence of two genders within one person. Proust believes that the invert is a woman in a man's body, a man who is actually a woman. From this conception he proceeds to a

62 See Darwin *Des différentes formes de fleurs dans les plantes de la même espèce*, Introduction, p. 10, and Chapter 7, p. 287.

genealogy of hermaphroditism through a personal reinterpretation of Aristophanes' myth narrated in the *Symposium*.[63] In fact, according to Proust, the original hermaphrodite (ancestor of the homosexual) consists of two different sexes. In Plato, on the other hand, woman attracted to other women comes from Zeus's separation of the woman consisting of two women, just as the man who feels attraction to other men from the separation of a being consisting of two men, while heterosexuals have their origin in the separation of the androgyne. Proust actually turns the myth upside down by introducing the idea – through Darwin – of an original, ancestral hermaphroditism, evidenced by the presence of traces of male organs in the anatomy of women and female organs in that of men.

Moreover, classical references will return, again with regard to homosexuality and M. de Charlus, in *La Prisonnière*, where the narrator highlights the difference between ancient homosexuality and that of his time: in Greece and Rome – in the texts of Plato, Theocritus and Virgil – a custom emerges, a way of life perfectly inscribed in the social canons of the time; today's homosexuality which 'survives and increases is only the involuntary [l'involontaire], the neurotic kind [la nerveuse]'; it is that 'which we conceal from other people and disguise to ourselves [celle qu'on cache aux autres et qu'on travestit à soi-même]'.[64] This is 'the homosexuality that survives in spite of obstacles, a thing of scorn and loathing, that is the only true form, the only form that can be found conjoined in a person with an enhancement of his moral qualities [l'homosexualité survivante malgré les obstacles, honteuse, flétrie, qui est la seule vraie, la seule à laquelle puisse correspondre chez le même être un affinement des qualités morales]'.[65]

This moral enhancement seems almost to evoke a humanistic declination of homosexuality.

---

63 Although Proust is familiar with *The Symposium*, he seems to reject the crucial implications of Aristophanes' sexual characterology. See Melvin Seiden, 'Proust's Marcel and Saint-Loup: Inversion Reconsidered', *Contemporary Literature*, 10.2 (1969), 220–40 (p. 229).

64 *Recherche*, III, p. 710; *SLT*, vol. 5: *The Captive and The Fugitive* ed. by William C. Carter, trans. by C. K. Scott Moncrieff (New Haven and London: Yale University Press, 2023), p. 219.

65 Ibid.

Armando Mascolo*

'NULLUM HOMINEM A ME ALIENUM PUTO'
The Agonic Humanism of Miguel de Unamuno

Abstract

For Unamuno, true humanism cannot rest on an abstract idea of humanity but must affirm the concrete individual, the man of flesh and bone. To affirm man is to affirm his personal, 'agonic' consciousness: the inner struggle between rational scepticism and vital feeling. Man is torn between reason, which reveals the abyss of his finitude and mortality, and life, which drives his longing for immortality. From this irreconcilable conflict arises the 'tragic sense of life', the very condition of philosophizing, in which the concrete man is both subject and object.

Keywords

Tragic sense of life – Humanism – Agonic consciousness – Suffering – Death

1. *An Incarnated Humanism: The Centrality of the 'hombre de carne y hueso'*

I want to make it clear right from the start the first part of the title of this essay, namely the Latin phrase *Nullum hominem a me alienum puto*, referring to the opening lines of the first chapter of what is certainly Unamuno's most famous and significant work, *Del sentimiento trágico de la vida*. At the beginning of this book, the

* Istituto per la Storia del Pensiero Filosofico e Scientifico Moderno – CNR.

Spanish philosopher defines the main object of his inquiry, clearly outlining the fundamental theoretical axis that characterizes his humanism, understood as a reflection on the concrete human being considered in his existential condition. Unamuno begins by quoting a Latin maxim from the Roman playwright Publius Terentius Afer, taken from *Heautontimorumenos* (*The Self-Tormentor*, 163 B.C., v. 77): '*Homo sum; nihil humani a me alienum puto*' ('I am a man; nothing human is alien to me'). This maxim enjoyed great popularity during the Humanist and Renaissance periods, becoming, in a certain sense, the emblem of the ideal of *humanitas* developed by the intellectuals of the time. From their perspective, the emphasis was placed on the shared belonging to the human condition, on the universal dignity of man, and on the possibility of an empathetic and compassionate understanding of his inner life. Unamuno, however, intervenes in a significant way on Terence's phrase, introducing a modification that, although seemingly minor, is in fact laden with theoretical implications. He replaces the adjective *humani* with the noun *hominem*, and thus writes:

> *Homo sum; nihil humani a me alienum puto*, said the Latin playwright. And I would rather say, *Nullum hominem a me alienum puto*: I am a man; no other man do I deem a stranger. For to me the adjective *humanus* is no less suspect than its abstract substantive *humanitas*, humanity. Neither the human nor humanity, neither the simple adjective nor the substantivized adjective, but the concrete substantive – man. The man of flesh and bone; the man who is born, suffers, and dies – above all, who dies; the man who eats and drinks and plays and sleeps and thinks and wills; the man who is seen and heard; the brother, the real brother.
>
> For there is another thing which is also called man, and he is the subject of not a few lucubrations, more or less scientific. He is the legendary featherless biped, the *ζῷον πολιτικόν* of Aristotle, the social contractor of Rousseau, the *homo economicus* of the Manchester school, the *homo sapiens* of Linnæus, or, if you like, the vertical mammal. A man neither of here nor there, neither of this age nor of another, who has neither sex nor country, who is, in brief, merely an idea. That is to say, a no-man.[1]

1 Miguel de Unamuno, *Del sentimiento trágico de la vida en los hombres y en los pueblos* (1913), in Id., *Obras completas*, 9 vols, ed. by Manuel García Blanco (Madrid: Escelicer, 1966–1971), vol. 7, p. 109. For the English translation, we have made use of the following text: Miguel de

The lexical shift is not without significance: what is being rejected is the generic reference to the human as an abstract category, in favour of the affirmation of man in his concrete individuality. As Unamuno himself makes clear, it is the use of the adjective *humanus* and the abstract noun *humanitas* that arouses suspicion, as they are expressions of a kind of thought that distances itself from the real experience of the human being, reducing it to an idea, a type, or a function.

The opposition that thus emerges is sharp and unmistakable: on the one hand, the empirical, historical, incarnate man, the one who lives, suffers, and dies; on the other, a conceptual figure, stripped of its existential determinations: the Platonic 'featherless biped', the Aristotelian *ζῷον πολιτικόν*, the *homo oeconomicus* of liberal theory, the *homo sapiens* of scientific taxonomy.

Unamuno therefore decisively distances himself from an abstract idea of humanity developed in particular by the scientific rationality typical of 19th-century positivist thought, setting against it the conception of a humanism centred on the concrete man, who is both the subject and, at the same time, the foremost object of all philosophy:

> [...] when a man affirms his 'I', his personal consciousness, he affirms man, man concrete and real, affirms the true humanism – the humanism of man, not of the things of man – and in affirming man he affirms consciousness. For the only consciousness of which we have consciousness is that of man.[2]

To affirm the concrete man, therefore, means at the same time to affirm his personal consciousness, which for Unamuno is an 'agonic' consciousness, that is, a consciousness of the 'struggle'[3] inherent in every human being between reason and life, or rather, between

---

Unamuno, *The Tragic Sense of Life in Men and in Peoples*, trans. by John Ernest Crawford Flitch, with an introductory essay by Salvador de Madariaga (London: Macmillan and Company, 1926), p. 1.

2 Ibid., p. 116; English trans., p. 13.

3 The term *agony* is used by Unamuno in its original etymological sense, derived from Ancient Greek: *ἀγωνία*, from *ἀγών*, meaning contest, struggle, or combat.

rational scepticism and vital feeling. For life, as Spinoza taught, is *conatus essendi*, a striving, an impulse towards the perpetuation of the self, against every contrary lesson of reason.

Here lies the central core of Unamuno's agonic humanism, which holds that human nature is marked by contradiction, since man is a divided, torn, agonic being, intrinsically split between the yearning for a sense of eternity and the finite reality of his earthly existence. From this perspective, life becomes a permanent war, a painful struggle, a continuous tension between feeling and reason, between the longing for immortality and the awareness of one's own finitude.

It is within this dramatic dialectic that Unamuno's agonic humanism takes root: a humanism that does not seek the reconciliation of opposites, but rather fully embraces their tension, making it the essential hallmark of the human condition.

2. *The reception of Darwinism in Spain*

The story of the spread of Darwin's ideas in Spain is rather complex and deeply influenced by the cultural, philosophical, and political climate of the second half of the nineteenth century.[4] Although Spanish naturalists had already become aware of the publication of *On the Origin of Species* as early as 1860, just a year after its release in England, direct access to the text by the Iberian public came several years later: the first complete translation of the work did not appear until 1877. Such a delay bears witness to the

4 For a more in-depth examination of the reception of Darwinism in Spain, see the following works: Diego Núñez Ruiz, *El darwinismo en España* (Madrid: Castalia, 1977); Thomas F. Glick, *Darwin en España* (Barcelona: Península, 1982); Xosé Antón Fraga, *La recepción del darwinismo por los naturalistas españoles del s. XIX, un análisis general*, in *Evolucionismo y cultura: darwinismo en Europa e Iberoamérica*, ed. by Miguel Ángel Puig Samper, Rosaura Ruiz Gutiérrez and Andrés Galera Gómez (Junta de Extremadura: UNAM y Doce Calles, 2002), pp. 249–67; Francisco Pelayo López, *Darwinismo y antidarwinismo en España (1900–1939)*, in *Evolucionismo y cultura*, pp. 267–84; Francisco Blázquez Paniagua, 'Breve historia del darwinismo en España', *AmbioCiencias: revista de divulgación*, 1 (2009), 23–9.

obstacles that hindered the entry of evolutionism into the Spanish cultural landscape, where Darwinian ideas were, for a long time, the subject of a debate that was more ideological than strictly scientific, becoming a genuine point of contention in which epistemological, ethical, and theological issues were interwoven.

A crucial moment for the spread of Darwinian thought in Spain was the period of the '*Sexenio Revolucionario*' (1868–1874), characterised by a more open political climate and, above all, by the abolition of censorship. This context allowed, for the first time, a public and relatively free discussion of evolutionism and its implications, even though such debate, as we have already seen, initially took place in the absence of complete translations of Darwin's principal works.

In this context, the decisive influence of European scientific cultures – particularly the French and German ones – played a crucial mediating role in the introduction of evolutionary ideas into Spain. In the absence of a solid domestic scientific programme capable of organically developing Darwin's thought, its reception took place through philosophical intermediaries, most notably Krausism, a current of idealist and rationalist inspiration that became an important vehicle for intellectual modernisation.

Krausism was a philosophical and cultural movement that, while spreading widely in Belgium, the Netherlands and Latin America, reached its height in Spain during the second half of the nineteenth century thanks to Julián Sanz del Río (1814–1869).[5] Through a series of works and translations – among which his *Ideal de la humanidad para la vida* (1860) certainly stands out – he played a major role in introducing into the Iberian Peninsula the doctrines of the German philosopher Karl Christian Friedrich Krause (1781–1832), from whom the movement drew its inspiration. Educated at the University of Jena, where he attended the lectures of Hegel, Fichte and Schelling, and later teaching at the universities of Berlin,

5 For a detailed examination of Sanz del Río's intellectual itinerary, see in particular the work of Fernando Martín, José Luis Abellán, *El krausismo español desde dentro: Sanz del Río, autobiografía de intimidad* (Madrid: Tecnos, 1978).

Munich and Göttingen, Krause is best known for having devised a philosophical system known as 'panentheism' or 'harmonic rationalism'. According to this system, the finite world is a reflection of the infinite, and the individual self represents a stage in a process of spiritual elevation and participation in the essence of the Absolute. Krause's philosophical doctrine thus sought to reconcile theism and pantheism, the finite and the infinite, maintaining that God is present in all things without being identical with them, as He transcends them; in other words, that the world and God are united without confusion, distinct yet inseparable.[6]

In Spain, Krausism found fertile ground in the intellectual and political community of the time, taking on predominantly ethical and pedagogical rather than metaphysical features, aimed at a profound cultural renewal of Spanish society. From this point of view, many intellectuals, philosophers, poets, writers and politicians were influenced by it, directly or indirectly. One might think, for instance, of several leading figures of the so-called Generation of '98 – such as Ángel Ganivet, Miguel de Unamuno, Pío Baroja, and Antonio Machado – as well as of the regenerationist movement represented by Francisco Giner de los Ríos, Gumersindo de Azcárate, Nicolás Salmerón, and Joaquín Costa, founders, among other things, of the seminal *Institución Libre de Enseñanza*. Its influence also reached Ortega y Gasset and, finally, Joaquín Xirau. Despite their undeniable individual differences, all were united by a common ethical impulse and an equally strong pedagogical concern; thus, it may be said that Spanish Krausism focused chiefly on the human formation of man, on the holistic education of the individual, on a true *Bildung*.

Although Spanish Krausism did not constitute a form of philosophical Darwinism, it nevertheless facilitated the assimilation of certain aspects of evolutionism, purging it of its materialistic and mechanistic implications. From this perspective, Darwinism was interpreted as a manifestation of an evolving natural order,

6 For a thorough analysis of Krause's philosophy and writings, refer to the comprehensive book by Enrique Menéndez Ureña, *Krause, educador de la humanidad: una biografía* (Madrid: Universidad Pontificia Comillas, 2024).

compatible with a teleological and spiritual vision of the cosmos. It was, in essence, a form of 'conciliatory rationalism' aimed at integrating new scientific discoveries with philosophical and theological concerns.[7]

This original reinterpretation of Darwinism found expression in the activities of the *Sociedad Española de Historia Natural*, founded in 1871 by a group of Spanish naturalists. Its more progressive members showed themselves open to adopting Darwinian theories, though without developing a genuine biological research programme on evolutionism. Many of them were influenced by Krausism and sought a reconciliation between Darwinian thought and the ethical-philosophical concerns characteristic of this current. The absence of a serious scientific debate was therefore compensated for by a philosophical reworking that reflected the tension between modernity and tradition, science and spirituality, the autonomy of reason and the authority of religious dogma.

In the early twentieth century, Darwinism gradually assumed increasing significance, becoming a symbol of progress, the secularisation of culture, and the autonomy of scientific thought. The systematic publication of Darwin's works by the Sempere publishing house, together with the dissemination of the monistic theories of the German biologist and philosopher Ernst Haeckel (1834–1919) by Peregrín Casanova (1849–1919), professor of anatomy at the University of Valencia, helped to create a new intellectual climate, particularly within academic circles.

---

7 For an overview of Spanish Krausism, see the following studies: Juan López-Morillas, *El krausismo español: perfil de una aventura intelectual* (Madrid: Fondo de Cultura Económica, 1980); Adolfo Posada, *Breve historia del krausismo español* (Oviedo: Universidad de Oviedo, 1981); Elena de Jongh Rossel, *El krausismo y la generación de 1898* (Valencia: Albatros Hispanófila, 1985); Elías Díaz García, *La filosofía social del krausismo español* (Madrid: Debate, 1989); Antonio Jiménez García, *El krausismo y la Institución Libre de Enseñanza* (Madrid: Ediciones Pedagógicas, 2002); Julio Simó Ruescas, 'La *Naturphilosophie* en España: la recepción del evolucionismo en el entorno de la tradicíon Krausista', *Asclepio: Revista de historia de la medicina y de la ciencia*, 56 (2004), 197–222.

It was in this context that, in 1909, a group of medical students from the University of Valencia organised a tribute to Darwin on the centenary of his birth. The event was preceded by lively polemics in the press and marked by an intense ideological debate, demonstrating how evolutionism continued to provoke resistance not only of a religious nature but also of a cultural one. Several naturalists declined the invitation to participate, fearing the repercussions of too direct an involvement. Among those who did take part, however, was Unamuno, who offered a personal and rather unorthodox interpretation of evolutionism, one that diverged from Darwinism proper. Unamuno had previously translated several works by the English philosopher Herbert Spencer (1820–1903)[8] who, although influenced by Darwin, had developed an ethical-social conception of evolution based on the idea of the moral progress of humanity.

The event concluded with the publication of a commemorative volume edited by a group of young intellectuals from Lorca (Murcia),[9] in which Darwin was celebrated as a symbol of progress and scientific modernity. Yet even on this occasion, evolutionism was absorbed more as an ideological emblem than as a scientific paradigm. In this act of cultural appropriation, Darwin was more celebrated than understood, more invoked than studied, transformed into an emblem of a new age without his ideas ever being fully integrated into epistemological debate.

In conclusion, the reception of Darwinism in Spain took place in a delayed and fragmented manner, hindered both by linguistic barriers and by ideological and cultural resistance. Only slowly did evolutionary ideas manage to enter academic and public discourse, often through the mediation of philosophies that, in seeking to reconcile reason and faith, nature and spirit, scientific progress and moral order, softened or reinterpreted their original meaning. Far from being a straightforward transplantation of ideas, Spanish Darwinism was the outcome of a long and complex process of cultural negotiation, one that revealed,

---

8 On this aspect of Unamuno's work, see the article by Nazzareno Fioraso, *Unamuno, traductor de Spencer*, in *Pensar la traducción. La filosofía de camino entre las lenguas*, ed. by Antonio Gómez Ramos (Madrid: Universidad Carlos III de Madrid, 2014), pp. 221–230.

9 Various Authors (ed.), *Darwin* (Lorca: Imprenta 'La Tarde de Lorca', 1909).

through its contradictions and adaptations, the way in which a society grapples with the challenges of modern thought.

3. *The Darwinian concept of struggle for life*

Among the many elements of Darwinian thought assimilated by Unamuno,[10] the struggle for life occupies a central position, as it fits organically into the structure of his philosophy, which is oriented around the notion of the persistence of being. As we have already observed, the concept of *agonia* in Unamuno's philosophical reflection takes the form of a struggle – *lucha por la vida* – which clearly recalls the famous Darwinian notion of *struggle for life*. This connection, far from being marginal, is explicitly thematised by Unamuno himself in various parts of his works and receives significant development in two key essays: *El darwinismo* (1882) and the *Discurso pronunciado en el Paraninfo de la Universidad de Valencia* (1909), written on the occasion of the first centenary of Darwin's birth, organised by the *Academia médico-escolar* of Valencia.

In the 1882 essay, which represents Unamuno's first systematic engagement with the thought of the English naturalist, he interprets the *struggle for life* from the standpoint of classical evolutionism: a biological competition for survival based on the conquest of vital resources, in which only the strongest are destined to perpetuate their characteristics, thereby contributing to the perfection of the species. The Spanish philosopher writes:

> All animals struggle incessantly to obtain their food. The strongest is the one who prevails, and this strength may be found in its physical constitution, in some special organ, or in more intelligent faculties, which serve as so many defensive – and even offensive – weapons that can grant it superiority in the struggle for life. Consequently, the

10 For a more in-depth examination of Unamuno's reading of Darwinian evolutionism, see in particular the study by Peter G. Earle, 'El evolucionismo en el pensamiento de Unamuno', *Cuadernos de la Cátedra Miguel de Unamuno*, 14–15 (1964), 19–28.

> best-constituted animals are those that survive, and their advantages are transmitted through reproduction to their offspring, who continue the struggle and gradually attain greater perfection, thus giving rise to new species.[11]

This view is further developed in the principle according to which the development of a living being is proportional to the sophistication of the means by which it faces the struggle: 'From this it follows that the more perfected the weapons with which each animal is equipped for the struggle for existence, the greater the development it must attain'.[12]

In the *Discurso* of 1909, however, the interpretation of the *struggle for life* undergoes a significant evolution. In light of later readings – particularly of the *Biologische Probleme* by the German entomologist William Henry Rolph[13] – Unamuno reinterprets the notion of the struggle for life no longer as a mere expression of the instinct for self-preservation, akin to the Spinozian conception of being, nor as a manifestation of the instinct for the perpetuation of the species, linked to the biblical precept 'be fruitful and multiply', but rather as an expansive impulse, a will to transcend the limits of one's own being, to assert oneself radically in the world:

> *Lucha por la vida – Struggle for life* – this is what it is called, and it is supposed to arise from the combat of each individual and each species to preserve itself. There is already, in Spinoza's *Ethics*, a famous proposition – the seventh of Part Three – in which he declares that the essence of a being is nothing other than the effort by which it strives to persist in its own being. But in the Bible it is written: 'fruitful and multiply', not preserve yourselves! And, as Rolph very rightly observes in his biological studies, it is not growth and multiplication that demand more nourishment from animals, and thus drive them to struggle; rather, it is the tendency towards more nourishment – towards excess – that causes them to grow and multiply. The living being's impulse is not towards self-preservation, but towards self-surpassing,

11 Miguel de Unamuno, *El darwinismo*, in Id., *Prensa de Juventud* (Madrid: Edición de Elías Amézaga, 1995), p. 58.

12 Ibid., pp. 59–60.

13 William Henry Rolph, *Biologische Probleme: Zugleich als Versuch einer rationellen Ethik* (Leipzig: Engelmann, 1882).

> towards imposing itself, towards absorbing others. Wretched are the conservatives, those who remain on the defensive![14]

This approach makes it possible to grasp more deeply the ontology of the *hombre de carne y hueso*, understood not as a static being or abstract essence, but as a restless dynamism, an agonistic tension towards transcendence and the singularity of existence.

In this context, Unamuno praises Darwin not only for the speculative power of his hypotheses, but above all for having succeeded in embodying the theory of evolution in a concrete, empirically grounded form of research:

> It is not enough merely to conceive vast and fertile ideas, nor to create poems of abstract thought as profound as Hegel's *Logic*; it is also necessary to bring them down to earth, to show them embodied in deeds, to give them solid ground. And although the vast doctrine of Evolution had been conceived, developed, and formulated before Darwin, he was the first to test it in a concrete field of universal life. The laws of the struggle for existence, of the survival of the fittest, of adaptation to the environment, of inheritance – it was Darwin who demonstrated them in practice. And above all, the law of selection. To him more than to anyone else is owed the principle of selection.[15]

Unamuno recognises in particular the debt that Darwin owed to the economic thought of Thomas Robert Malthus, from which his conception of the vital struggle originated:

> When Darwin made his voyage on the Beagle, he had already conceived a provisional hypothesis while studying the geographical distribution of fauna and flora; but it was in 1838, upon reading the economist Malthus, that he received the sudden flash of insight which illuminated his speculations. From Malthus he learned of the struggle for existence, arising from the fact that resources tend to increase in arithmetic progression, while living beings tend to increase in arithmetic progression. The ideas this suggested to him, he sought to confirm empirically by studying cases of artificial selection carried out

14 Miguel de Unamuno, *Discurso pronunciado en el Paraninfo de la Universidad de Valencia*, in Id., *Obras completas*, vol. 9, p. 254.

15 Miguel de Unamuno, *Darwin*, in Id., *Obras Completas*, vol. 8, p. 200.

by breeders, and then by examining the geographical distribution of animals and plants, as well as the study of palaeontology.[16]

To the concept of the *struggle for life*, Unamuno adds, in a complementary manner, that of spontaneous variation, a crucial element for understanding the individual and unrepeatable dimension of being. Although not a recurring theme in his work, variation – understood as ontological novelty – legitimises the principle of individuation, a cornerstone of the philosophy of existence:

> Alongside this struggle – not merely for existence, but for over-existence, for domination – there exists, and in order that the struggle may exert its effect, what Darwin called the tendency towards spontaneous variation: the fact that each new living individual is truly new, differing in some small or significant way from those that came before, born with some peculiarity. It has been said that no two leaves of a tree are identical.[17]

From this perspective, the enigma of variation – which Darwin himself acknowledged – becomes, for Unamuno, an opportunity to propose an original philosophical response, influenced particularly by the interpretation of Darwinism offered by Eduard von Hartmann[18] and Rudolf Albert von Kölliker.[19] In the essay *Evolución y revolución*, published in March 1889 in *El Noticiero Bilbaíno*,

---

16 Unamuno, *Discurso*, p. 253.

17 Ibid., p. 255.

18 In 1875, Eduard von Hartmann, the pessimist philosopher known for his *Philosophie des Unbewussten* (1869), wrote *Wahrheit Und Irrthum Im Darwinismus* (Berlin: Carl Duncker's Verlag, 1875), in which he offers a critical analysis of Darwin's theory of evolution. A few years later, in 1879, the work was translated into Spanish under the title *El darwinismo: lo verdadero y lo falso de esta teoría* (Sevilla/Madrid: Imprenta de R. Baldaraque, 1879), and it is not unlikely that Unamuno read this very version, although he was perfectly proficient in German.

19 Rudolf Albert von Kölliker, Swiss anatomist and physiologist, is considered the German father of 19th-century biology. In 1864, he published *Über die Darwin'sche Schöpfungstheorie*, in which, dealing with *On the Origin of Species*, he criticised Darwinian doctrines as being overly teleological, asserting that certain traits appear suddenly rather than through small incremental steps.

Unamuno considers the Darwinian concepts of inheritance and selection to constitute the foundation of the doctrine of evolution, thereby interpreting Darwinian evolution as a slow and gradual transformation, in which species pass from one form to another without abrupt changes, due to the action of external circumstances on what has been inherited. However, finding the Leibnizian assumption that *natura non facit saltus*, on which Darwin's evolutionism is based, not particularly convincing, he aligns himself with the school of heterogeneous generation of Kölliker and Hartmann, which, according to his interpretation, completes Darwinian evolutionism with the concept of 'vital revolution': a constitutive rupture that reveals the irreducibility of the individual and their vocation for transcendence.[20]

4. *Man as an agonic being: suffering and meditatio mortis*

In Unamuno, as we have seen, human life is configured as a permanent struggle between feeling and reason, between the anxiety for immortality and the awareness of one's own finitude. Given this constant tension in man between reason and life, suffering becomes the most immediate revelation of consciousness. In the Spanish philosopher, suffering rises to the rank of an ontological category and serves as the supreme criterion of existence:

> Suffering is the path of consciousness, and by it living beings arrive at the possession of self-consciousness. For to possess consciousness of oneself, to possess personality, is to know oneself and to feel oneself distinct from other beings, and this feeling of distinction is only reached through an act of collision, through suffering more or less severe, through the sense of one's own limits. Consciousness of oneself is simply consciousness of one's own limitation.[21]

---

20 Cf. Miguel de Unamuno, *Evolución y revolución*, in Id., *Unamuno y el socialismo. Artículos recuperados (1886–1928)*, ed. by Diego Núñez and Pedro Ribas (Granada: Editorial Comares, 1997), pp. 75–83.

21 Unamuno, *Del sentimiento trágico de la vida*, p. 192; English trans., p. 140.

Awareness of one's own limitations, which is awareness of one's own mortality. We have seen how at the centre of Unamuno's humanism lies the concrete man, 'of flesh and bone; the man who is born, suffers, and dies – above all, who dies',[22] Unamuno is careful to stress. For it is precisely death that, according to the Spanish philosopher, tragically marks the human condition. 'The overall Spanish sense of life – wrote the italian philosopher Lorenzo Giusso in his remarkable essay on Don Quixote – can be identified with that expressed so brilliantly by Calderón in the auto sacramental *El gran teatro del mundo*, as in the famous drama *La vida es sueño*: a sense of impending vanity, of inevitable death, of the spectral insubstantiality of our destiny'.[23]

A deep-seated unease indeed seems to run through Spanish philosophical reflection from Seneca to its most significant manifestations in the twentieth century, an unease rooted in the dramatic awareness of the transience of human nature. And if there is one figure in the Spanish philosophical landscape who more than any other has reflected on the theme of death, making it his 'secret obsession', it is undoubtedly Unamuno.

The centrality of *meditatio mortis* in Unamuno does not carry a merely psychological significance, 'because it is precisely in the notion of death that the spiritual puberty is connected, in which peoples, like individuals, grow until they reach what the Spanish philosopher defines as the tragic sense of life'.[24] At the origin of Unamuno's reflection on death — which coincides with the philosopher's most mature thought, marked by the publication of his most significant works such as *Vida de Don Quijote y Sancho*, *Del sentimiento trágico de la vida*, and *La agonía del Cristianismo* – lies his personal experience of disorientation, of inner conflict, stemming from the profound spiritual and philosophical crisis he

22 Ibid., p. 109; English trans., p. 1.

23 Lorenzo Giusso, *Il mito di Don Chisciotte 5* (13 novembre 1956), in Id., *Autoritratto spagnolo*, ed. by Alberto Spaini (Turin: ERI–Edizioni Rai, 1959), p. 244.

24 Ferruccio Masini, 'Filosofia della morte in Miguel de Unamuno', *Cuadernos de la Cátedra Miguel de Unamuno*, 8 (1958), 27–42 (p. 27).

underwent in March 1897.[25] This was likely triggered by a nocturnal attack of angina pectoris, which forced him to confront the idea of death and whose repercussions flowed into the pages of the *Diario íntimo*, a tormented record of all his anxieties and abyssal fears.[26]

The *Diario íntimo*, the result of a long and uninterrupted gestation between 1897 and 1902, is the faithful record of what Unamuno himself called the experience of a 'mortal awakening', that is, the bitter awareness of the infinite vanity of all things, of the 'erosive force of time'[27] that overwhelms everything and constitutes the very root of the tragic sense of life, as it is inexorably bound to death. It is precisely because of this harsh awareness, arising from the crisis of 1897, that Unamuno would feel increasingly torn by an irreconcilable inner dissension, bearing upon himself the full weight of an irreducible dualism, a fracture between two opposing forces: on one side reason, which confronts him forcefully with the unbridgeable abyss of his own finitude, of his mortal being; on

25 Unamuno's religious crisis was first brought to light by the study of Antonio Sánchez Barbudo, 'La fe religiosa de Unamuno y su crisis de 1897', *Revista de la Universidad de Buenos Aires*, 18 (1950), 381–443. Regarding the significance of the crisis in the development of Unamuno's thought, scholars have often expressed divergent opinions. See, in this regard: Enrique Rivera de Ventosa, 'La crisis religiosa de Unamuno', *Cuadernos de la Cátedra Miguel de Unamuno*, 16–17 (1966), 107–133; Antonio Sánchez Barbudo, *Una experiencia decisiva: la crisis de 1897*, in Id. (coord.), *Miguel de Unamuno* (Madrid: Taurus, 1974), pp. 95–122; Paolo Tanganelli, *Unamuno fin de siglo. La escritura de la crisis* (Pisa: Edizioni Ets, 2003); Vincenzo Vitiello, 'Il cristianesimo tragico di Miguel de Unamuno. "No me mueve, mi Dios, para quererte/el cielo que me tienes prometido"', *Il Pensiero*, 2 (2006), 7–20; Armando F. Zubizarreta, *La inserción de Unamuno en el cristianismo: 1897*, in Id., *Tras las huellas de Unamuno* (Madrid: Taurus, 1960), pp. 111–51.

26 The five school notebooks that make up the *Diario íntimo* were discovered by Armando Zubizarreta in cupboard at the Rectorate of the University of Salamanca and were first published in Spain in 1970. Cf. Miguel de Unamuno, *Diario íntimo*, ed. by E. Kerrigan, E. Naval, A. Froufe (Madrid: Alianza Editorial, 1970).

27 Pietro Prini, *Unamuno e la 'meditatio mortis'*, in Id., *Storia dell'esistenzialismo. Da Kierkegaard a oggi* (Rome: Studium, 1991), p. 75.

the other, life, which manifests itself through the desire to persist indefinitely, instilling in his heart the feeling of an inextinguishable 'anxiety for immortality'.

A careful and perceptive scholar of Spanish culture, Adriano Tilgher, rightly observed that the entire philosophical framework of Unamuno is rooted precisely 'in the fully lived experience of the radical antinomy of reason and life, which from Kant onwards is one of the central motifs of modern speculation'.[28] From this 'radical antinomy' between reason and life arises what Unamuno calls the tragic sense of life, which constitutes the very condition of philosophising, since

> the tragic history of human thought is simply the history of a struggle between reason and life – reason bent on rationalizing life and forcing it to submit to the inevitable, to mortality; life bent on vitalizing reason and forcing it to serve as a support for its own vital desires.[29]

Both subject and object of this philosophising is the concrete man, 'de carne y hueso', a man concerned with his personal destiny, who questions himself, who wonders what will become of him, deeply torn between the unfulfilled desire to be immortal and the anguishing thought of nothingness and death to which reason constantly returns him. In this continuous and extreme dialectical tension between life and reason, no Hegelian synthesis is possible: the opposition between these two antagonistic forces is irresolvable and constitutes the 'intranscendable' condition of our very existence, destined to unfold 'agonically' in a struggle and incessant confrontation between reason and vital feeling. On this point, Unamuno writes:

> For living is one thing and knowing is another; […] perhaps there is such an opposition between the two that we may say that everything

28 Adriano Tilgher, *La visione della vita di Miguel de Unamuno*, in Miguel de Unamuno, *La Sfinge senza Edipo*, trans. by Piero Pillepich (Milan: Corbaccio, 1925), p. 10.

29 Unamuno, *Del sentimiento trágico de la vida*, p. 177; English trans., p. 115.

> vital is anti-rational, not merely irrational, and that everything rational is anti-vital. And this is the basis of the tragic sense of life.[30]

Unamuno's man positions himself 'agonically' as the tragic 'meeting point' between life and reason, between the anxiety for the infinite and the vertigo of his own finitude, since the despair of vital feeling and the frustration of rational scepticism cast him into the 'depths of the abyss' of his own consciousness, there where these two antagonistic forces confront each other to embrace tragically, an embrace that entails an insurmountable tension, a perpetual struggle with neither victory nor hope. Unamuno writes:

> The vital longing for human immortality finds no consolation in reason and that reason leaves us without incentive or consolation in life and life itself without real finality. But here, in the depths of the abyss, the despair of the heart and of the will and the scepticism of reason meet face to face and embrace like brothers. And we shall see it is from this embrace, a tragic – that is to say, an intimately loving – embrace, that the wellspring of life will flow, a life serious and terrible.[31]

The vital feeling alone would lead man to reassuring certainties of immortality, while reason alone would lead to the certainty of nothingness. Both outcomes prove wholly inadequate to the profound complexity that characterises the human condition, to which the tragic sense of life is inherent, and therefore a state of agonising despair. It should be noted, however, that this despair does not correspond to the 'mortal sickness' overcome through the leap of faith described by Kierkegaard,[32] but rather to a permanent condition of human existence. From this despair arises that feeling of 'incertitude' which Unamuno defines as 'saving', because it is what makes faith possible, however always a risky faith, in our immortality:

---

30 Ibid., p. 129; English trans., p. 34.

31 Ibid., p. 172; English trans., p. 106.

32 For the significant presence of Kierkegaard in Unamuno's thought, see in particular the substantial study by Jesús Antonio Collado, *Kierkegaard y Unamuno. La existencia religiosa* (Madrid: Editorial Gredos, 1962).

> Reason does not lead me and cannot lead me to doubt that I exist. Whither reason does lead me is to vital scepticism, or more properly, to vital negation – not merely to doubt, but to deny, that my consciousness survives my death. Scepticism is produced by the clash between reason and desire. And from this clash, from this embrace between despair and scepticism, is born that holy, that sweet, that saving incertitude, which is our supreme consolation.[33]

The dramatic experience of death, in its dark unpredictability, therefore does not lead Unamuno to the peace of resignation, but rather existentially commits him to struggle. From this arises, as a kind of rebellious opposition, life understood as a perpetual struggle – indeed, one might even say a desperate struggle – since it is a fight driven by itself and not by the certainty of ultimate victory. This struggle manifests in the contradictory essence of human existence, torn by the irreconcilable conflict between heart and mind, between feeling and reason. For this reason, the 'existential' or 'agonic' man – that is, simply and intensely human – is he who suffers and lives through this intimate dissension between the life of reason and the reasons of life.

Having reached the bottom of the abyss of his own consciousness, the theatre of the irreconcilable conflict between reason and vital feeling, Unamuno therefore asserts the necessity to 'accept the conflict as such and to live by it',[34] so that this 'abyss of despair' may become the source 'of a vigorous life, of an efficacious activity, of an ethic, of an esthetic, of a religion and even of a logic',[35] which is nothing other than the 'logic of the heart', demonstrating how a certain 'transcendental pessimism' is capable of generating a 'temporal and terrestrial optimism'.[36] It is therefore necessary

33 Unamuno, *Del sentimiento trágico de la vida*, p. 179; English trans., p. 118.

34 Ibid., p. 183; English trans., p. 124.

35 Ibid.

36 Ibid., p. 187; English trans., pp. 130–1. On the concept of 'transcendental pessimism', see above all the important monograph by Pedro Cerezo Galán, *Las máscaras de lo trágico. Filosofía y tragedia en Miguel de Unamuno*, prólogo de Pedro Laín Entralgo (Madrid: Editorial Trotta, 1996), pp. 273–310.

to overcome every kind of 'passive pessimism' in the name of what Unamuno, in his essay entitled *¡Adentro!*, defined as 'active resignation',[37] recognising in the tragic sense of life 'the spring of heroic achievements',[38] to the extent that it is possible to assert that 'many of the greatest heroes, perhaps the greatest of all, have been men of despair and that by despair they have accomplished their mighty works'.[39]

Unamuno's tragic agonism is thus permeated by a sort of 'heroic spirit', or rather, by a 'heroic will'[40] that makes literary creation an adventure and a struggle to affirm one's individuality against death and nothingness. Unamuno's transcendent pessimism, therefore, 'lives the heroic struggle to transcend oneself creatively in the form of ethical *engagement*', drawing on 'the unconditional affirmation of consciousness as a will to be'.[41]

From this perspective, the figure of Don Quixote, so central to Unamuno's thought, can be understood as the 'fantastic universal'[42] of the desperate hero, as the very symbol of an agonic humanism founded on a morality of heroic creation. Don Quixote – writes the Spanish philosopher – is ultimately nothing other than 'a man of an heroical

---

37 Miguel de Unamuno, *¡Adentro!* (1900), in Id., *Obras completas*, vol. 1, p. 949.

38 Unamuno, *Del sentimiento trágico de la vida*, p. 183; English trans., p. 125.

39 Ibid., pp. 186–87; English trans., p. 130.

40 Cf. Cerezo Galán, *Las máscaras de lo trágico*, pp. 311–71.

41 Armando Savignano, *Introduzione a Unamuno* (Rome–Bari: Laterza, 2001), p. 43.

42 In an article devoted to a comparison between Cervantes and Vico, Giuseppe Cacciatore argues that the figures of Don Quixote and Sancho can be understood as 'two extraordinary examples of the practical expression of Vico's theory of fantastic universals. On the one hand, madness, the generous utopia that moves between tradition and the future, the transfiguration of reality into hallucination and dream; on the other, the wisdom of daily toil, the realism of resigned acceptance of the present, but also the intelligence of surviving the hardships of the world'. Cf. Giuseppe Cacciatore, *Figure dell'ingegno in Cervantes e Vico*, in Id., *In dialogo con Vico. Ricerche, note, discussioni*, ed. by Manuela Sanna, Rosario Diana, Armando Mascolo (Rome: Edizioni di Storia e Letteratura, 2015), p. 231.

despair, the hero of the inward and resigned despair', and as such, his image must be elevated to an 'eternal exemplar of every man whose soul is the battle-ground of reason and immortal desire',[43] for it is only from despair that 'heroic hope, absurd hope, mad hope'[44] can be born.

Unamuno resolved, until the end, to take responsibility for his own despair by embodying the 'agonic idealism' of Don Quixote.[45] He committed himself to living tragically the irresolvable conflict between the anxiety for immortality and the abyss of nothingness, engaging in his personal battle against death while fully aware that he was destined for ultimate defeat – perhaps also strengthened by Schopenhauer's lesson that 'a happy life is impossible', for 'the highest that man can achieve is the heroic life'.[46] It is in this dramatic awareness, which knows no consolation, that, according to Unamuno, the true and deepest value of human existence is revealed.

43 Unamuno, *Del sentimiento trágico de la vida*, p. 180; English trans., p. 120.

44 Ibid., p. 299; English trans., p. 324. Of particular interest is the contrast that Pedro Laín Entralgo draws between the concept of hope in Unamuno and that of anguish in Heidegger. See, in this regard, the following works: Pedro Laín Entralgo, *Miguel de Unamuno o la desesperación esperanzada*, in Id., *La espera y la esperanza* (Madrid: Alianza Editorial, 1984), pp. 382–419; Pedro Laín Entralgo, *La esperanza de Unamuno*, in Id., *Esperanza en tiempo de crisis. Unamuno, Ortega, Jaspers, Bloch, Marañón, Heidegger, Zubiri, Sartre, Moltmann* (Barcelona: Círculo de Lectores, 1993), pp. 51–70.

45 On this specific aspect of Unamuno's thought, see above all the classic study by Michele Federico Sciacca, *Il chisciottismo tragico di Unamuno* (Milan: Marzorati, 1971).

46 Arthur Schopenhauer, *Parerga e paralipomena*, 2 vols (Milan: Adelphi, 2007), vol. 2, ed. by Mario Carpitella, trans. by Mazzino Montinari and Eva Amendola Kuhn, § 172a, p. 421.

François Specq*

UNSETTLING THE LAND
Henry David Thoreau and the Human-Nature Entanglement

Abstract

Henry David Thoreau (1817–1862), one of the foremost American writers of the nineteenth century, was both an ardent observer of nature and one of the most searching thinkers regarding the evolving sense of man's relation to environment. This essay specifically explores Thoreau's artful staging of competing modes of environmental consciousness in *Walden* (1854), and shows how this provides the groundwork for his pioneering call for nature preservation. Carefully framing his narrative so as to question standard ways of appropriating the world (scientific inquiry and economic exploitation of natural resources), and propounding a poetics in which the dynamics of writing (trans)figures the processes of nature, Thoreau grounded his environmental advocacy in a broader regard for the indomitable power and enduring secrecy of the physical world. Accepting the existence of nature as exceeding human control and signifying processes shifts the human/nonhuman divide, thus laying the ground for a more complex and balanced humanism.

Keywords

Thoreau – American Literature – Ecocriticism – Environmentalism – Nature Writing

* Ecole Normale Supérieure de Lyon.

1. *Introduction*

Writing in a highly charged period in the history of the new nation called the United States – characterized by territorial expansion, political strife, and deep cultural aspirations – the author and thinker Henry David Thoreau (1817–1862) offered a searching exploration of the link between the notion of humanism and that of nature, in so far as his conception of man's relation to the natural world involves a shift of the human/nonhuman divide. Indeed, while Thoreau has long been known as an early advocate of nature preservation, attention to this topic has mostly focused on his explicit calls in favor of wild nature, especially those dealing with the designation of specifically protected areas, such as national parks. My purpose, in this essay, is to outline Thoreau's more broadly-conceived environmentalism, which is more generally tied to an exploration of the various ways human beings relate to their environment, as part of his effort to foreground a more balanced form of humanism – one that is not premised on a sense of the superiority of mankind vis-à-vis a compliant natural world, but which takes stock of their deep interrelations or entanglement, while maintaining their separateness.

Thoreau's entire œuvre may be described as dealing with these issues. The present analysis will limit itself, however, to his best-known work, *Walden*—published in 1854. I hope to show that Thoreau's approach was both related to a specific context, and meant to have universal appeal. On the most simple level, *Walden* tells the story of Thoreau's two years' stay on the shores of Walden pond, near Boston, Massachusetts, in 1845–1847. Thoreau regarded this as an experiment, which allowed him to examine, and to question, the assumptions that had shaped the first few decades of the new nation—the United States was created in 1783, i.e., about 60 years before his venture. Thoreau examined the core tenets of the American national experience in multiple ways, which fundamentally revolve around the idea and project of settling the land. When Thoreau moved to a cabin he built in the woods near the lake of Walden, he was deliberately acting like a settler: making a new start, and aping settler life (playing settler, in a way). That is reflected in the narration, but also, visually, in the two images which appear in the book. First, an engraving of Thoreau's cabin at Walden, in the frontispiece, drawn by his sister, which looks like a pioneer's home on what was called 'the frontier' – images of which were popular in the nineteenth century (especially in the guise of homesteaders, after the Homestead Act of 1862, as celebrated in one of Currier & Ives's widely disseminated engravings, *Pioneer's Home* [1867]). Second, a map of the lake of Walden, inserted toward the end of the book, which was drawn by Thoreau himself. Taken together, the two images epitomized the archetypal American experience: settling the land, controling/framing it. (Other parts of the book explored related topics, such as the chapter on cultivating a bean-field.)

Thoreau's purpose was definitely not to illustrate pioneer life, however, but to interrogate and question the cultural and philosophical underpinnings of that act of settlement – especially its power to shape identities and communities, as well as its adherence to a simplistic humanist project of land transformation. As a result, his undertaking may broadly be defined as one of *un*settling, in both senses of the word, i.e.:

– the opposite of the historical settling of the continent (through parody – e.g. rejecting the characteristic idea of hard work);

– and the (more philosophical) effort at removing fixity (i.e., challenging standard assumptions or perspectives).

As a whole, Thoreau's book aimed at effecting emancipation – emancipation from a narrow understanding of the humanist project, if we broadly define the latter as imposing meaning on the world and transforming it. I will make it clear that his purpose was not to negate humanism, but to redefine it.

The frontispiece is thus highly symbolic of what is at stake: the deceptively simple image of Thoreau's cabin at Walden points to the book's raising the question of inhabitation, of dwelling in the world. Indeed, even as the physical world becomes a focus of attention in its own right, it is emphatically not taken for granted, but subjected to a powerful process of defamiliarization.

Although this is achieved throughout *Walden*, I will more particularly focus on the book's final two chapters, entitled 'The Pond in Winter' and 'Spring'. As their titles suggest, these complete the seasonal narrative initiated by Thoreau's settling by Walden Pond on July 4, 1845. I would like to suggest, however, that they actually were deliberately framed so as to offer more than mere seasonal narrative, but a sustained exploration of three opposite modes of environmental awareness: by this, I mean ways of envisioning the gap between matter and consciousness: Thoreau's charting of Walden Pond, and the harvest of the Walden ice, both from 'The Pond in Winter'; and the famous flowing sandbank passage from the 'Spring' chapter.

2. *Charting space: intellectual appropriation*

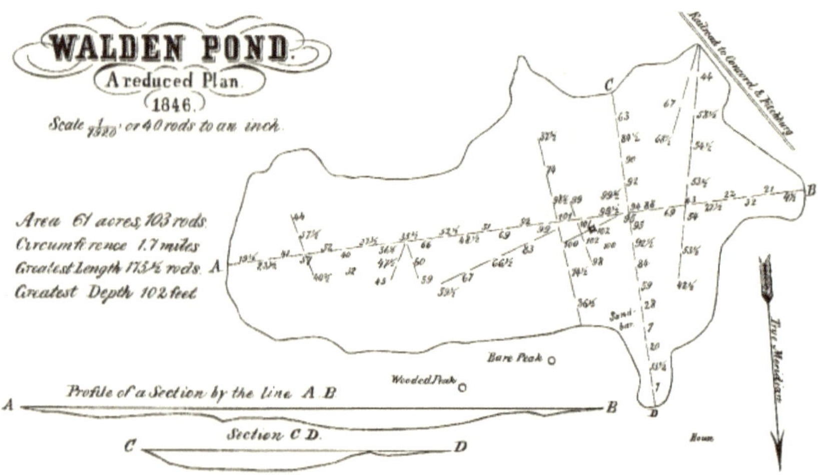

In the first sequence, Thoreau describes his drawing of a map of the lake on the shores of which he conducted his experiment. It is an event which really took place, and that was a serious undertaking (a form of business): indeed, Thoreau was a professional surveyor. In so doing, he was mimicking the stance which had shaped the settlement of the new nation, in which imposing order on the land through cartography was a crucial undertaking—in the aftermath of independence, for instance, the U.S. Congress's Land Ordinance of 1785 (which divided the Northwest Territory, West and North of the Ohio River, into neat one-mile squares) was a powerful tool for affirming the existence and power of the new nation, thus offering crucial support for domination of the land and territorial expansion, and the accompanying doctrine of Manifest Destiny.

From a more conceptual perspective, mapmaking illustrates the demands of rational inquiry that characterize modern science, and especially appears as an activity heeding Alexander von Humboldt's call for the 'delineation of nature'.[1] Indeed, *Walden* was framed as a response to Humboldt's *Cosmos*, which was translated into English in 1850.[2] Thoreau's desire to write as if 'from a distant land'[3], for instance, echoed Humboldt's notion of the equivalence between experiment and the infinity of the world, which is itself a powerful defamiliarizing factor: 'The study of a science that promises to lead us through the vast range of creation may be compared to a journey in a far-distant land'[4]. By drawing a map, Thoreau was also performing the task set by Humboldt of 'trac[ing] the *stable* amid the vacillating, ever-recurring alternation of physical metamorphoses'.[5] However playfully, mapmaking seems to abolish the mutability of things – to settle them – thus enforcing or supporting a rhetoric of the ideal, which in its turn can uphold a rhetoric of empire, as suggested by Humboldt:

1 Alexander von Humboldt, *Cosmos: A Sketch of a Physical Description of the Universe*, trans. by Elise C. Otté, 2 vols (New York: Harper, 1850), Vol. I, p. 79.

2 As Laura Dassow Walls notes, 'Thoreau's local would always speak to the cosmic: *Walden*, like [Edgar Allan Poe's] *Eureka*, was a response to Humboldt's *Cosmos*'. Laura Dassow Walls, *Seeing New Worlds: Henry David Thoreau and Nineteenth-Century Natural Science* (Madison: The University of Wisconsin Press, 1995), p. 264. Walls has offered the most sustained exploration of the importance of Alexander von Humboldt in nineteenth-century American culture generally (*The Passage to Cosmos: Alexander von Humboldt and the Shaping of America*, Chicago: The University of Chicago Press, 2009), and in Thoreau's thinking specifically (*Seeing New Worlds*).

3 Henry David Thoreau, *Walden* (1854), ed. by James Lyndon Shanley (Princeton: Princeton University Press, 1971), p. 3.

4 Humboldt, *Cosmos*, I, p. 50.

5 Humboldt, *Cosmos*, I, p. xii (emphasis in original).

> [...] so ought we likewise, in our pursuit of science, to strive after a knowledge of the laws and the principles of unity that pervade the vital forces of the universe; and it is by such a course that physical studies may be made subservient to the progress of industry, which is *a conquest of mind over matter*.[6]

Within that tradition, the map represents the synthesizing power of knowledge. It points to an experience of the physical world as mediated by the power of reason – as advocated by Humboldt: 'the traveler [...] is guided by reason in his researches'.[7] Mapmaking, as a distinctive form of humanistic knowledge, was meant to disprove popular assumptions – in this case, about the lake's bottomlessness ('It is remarkable how long men will believe in the bottomlessness of a pond without taking the trouble to sound it'[8]).

Mapmaking thus represents or symbolizes a form of intellectual appropriation of the land. Indeed, it enacts a characteristically dualistic relation to nature, premised on a clear distinction between the material world and consciousness, with a view to producing or increasing intelligibility – the map represents the clarifying and synthesizing power of knowledge. In the rhetorical structure of the latter part of *Walden*, the mapmaking sequence plays an essential role as effecting an objectification of the lake, standing for the different ways human beings use a range of intellectual frameworks to make sense of the world, to understand, circumscribe, and encompass it: in this process the world becomes an object of rational knowledge, undergoing a mighty transformation by being experienced as a site of measurement rather than imagination, which is here dismissed as fancy: 'The amount of it is, the imagination, give it the least license, dives deeper and soars higher than Nature goes', the narrator notes.[9] Through this process of *imaging* – as opposed to imagining – the otherness of nature is denied, or at least reduced, as it is bent to the framing power of reason – in other words, 'translated'. Significantly, Thoreau is able to draw a kind of metaphorical lesson from his

6 Humboldt, *Cosmos*, I, pp. 53–4 (my emphasis).
7 Humboldt, *Cosmos*, I, p. 51.
8 Thoreau, *Walden*, p. 285.
9 Thoreau, *Walden*, p. 288.

experience: 'What I have observed of the pond is no less true in ethics', he writes.[10] But, no less significantly, he gives that ethical translation a turn which adumbrates what he later says about the limits of science vis-à-vis poetic perception:

> It is true, we are such poor navigators that our thoughts, for the most part, stand off and on upon a harborless coast, are conversant only with the bights of the bays of poesy, or steer for the public ports of entry, and go into the dry docks of science, where they merely refit for this world, and no natural currents concur to individualize them.[11]

To conclude regarding this first sequence, one should also add that we can already sense the subtle twist Thoreau gave to the standard national narrative: the strikingly irregular shape of the outlying, asymmetrical body of water called Walden Pond, may certainly be seen as an oblique comment on the American passion for the conquest and dehumanization of space through geometry – a stance that reaches back to William Penn's gridiron plan for Philadelphia (1681), and, in an even more spectacular way, to the U.S. Congress's Land Ordinance of 1785, which enabled surveyors to impose a grid pattern on the landscape. Thoreau's map of Walden Pond – along with his emphasis on winding paths rather than straight lines – implicitly questioned the way of relating to the land underlying the logic of territorial expansion.

3. *Exploiting natural resources: material appropriation*

The second sequence considers what is even more central to the idea of settling the land, i.e., the exploitation of natural resources. Thoreau thus moves on to depict a more literal appropriation of the land, one conducted in the name of economic materialism: the harvesting of the Walden ice illustrates the wide-ranging exploitation of natural resources for a profit. This is also a real, historical event: the winter following the pond survey, a crew of a hundred ice-cutters

10 Thoreau, *Walden*, p. 291.
11 Thoreau, *Walden*, p. 292.

arrived at Walden to cut the ice for shipment to various places all over the world, including India or Brazil for instance – that was before the invention of the refrigerator, and an early example of globalization! The whole process was closely linked to the development of railroads, which is indeed another important theme in *Walden*. That epitome of extractivist capitalism was an international, highly profitable industry. Ice-harvesting was then an ordinary event, which was widely documented and depicted. Thoreau defamiliarized it through close description and mythologizing – that topic was certainly not a common thing for literature to include at the time.

Considered as part of a carefully-organized three-stage structure, ice-cutting parallels mapmaking to some extent, insofar as both are socially acceptable undertakings, aspiring to social approbation and – intellectual or commercial – gain: a form of *business*, as opposed to the *idleness* of the wandering writer, which characterizes the third sequence. However, the logging of the ice also significantly differs from the charting of the lake, being very serious business indeed, and totally at odds with any form of inquiry or exploration. Here matter is a mere resource to be exploited for practical purposes and material benefit. Even more than mapmaking, ice-cutting enforces the power of measure, linearity, sequence, and, ultimately, containment.

Thoreau's point, in the philosophical framework of *Walden*, is that the logging of the ice is not liable to any form of translation or conversion, other than monetary: ice-cutting intrinsically negates any form of intellectual or metaphoric substitution, but instead favors repetition of the same – as reflected by the telling image of the contractor commissioning the exploitation of the Walden ice 'in order to cover each one of his dollars with another',[12] or the image of the ice blocks 'placed evenly side by side, and row upon row':[13] instead of producing intellectual difference or poetic expansion, economic exploitation contracts and reduces the real. The only substitution it operates is of stasis for process and energy and (seasonal) flux: the monumental, static icestack is meant to subtract matter from its normal life cycle, and to obliterate the

12 Thoreau, *Walden*, p. 294.
13 Thoreau, *Walden*, p. 295.

ordinary succession of seasons, by providing cold in the summer. Ice-cutting thus epitomizes the capitalistic downplaying of the present in favor of future interest. As Thoreau depicts a relation to nature governed by the logic of the commodity, and exposes the exploitative character of settler colonialism, meant to reap a profit, he draws attention to the material containment of the physical world. That sequence illustrates his sharp critique of the United States as a land of opportunity, in which technology, self-interest, commodity fetishism, accumulation, and monetary value prevail, turning the country into a land of expediency. Beyond the specific critique of the United States, however, Thoreau meant to convey the idea of the misguided nature of the standard, narrowly-conceived humanistic project, based on an appropriative relation to the world.

4. *Nature's independent powers: unsettling the land*

Thoreau's powerful depiction of the economic subjection of the world is meant to set the stage for the contrasting logic operating in the third sequence, significantly taking place in the Spring, with its associated ideas of thawing and rebirth. The thawing bankside section of the 'Spring' chapter is one of Thoreau's best-known pages:

> Few phenomena gave me more delight than to observe the forms which thawing sand and clay assume in flowing down the sides of a deep cut on the railroad [...]. The material was sand of every degree of fineness and of various rich colors, commonly mixed with a little clay. When the frost comes out in the spring [...] the sand begins to flow down the slopes like lava, sometimes bursting out through the snow and overflowing it where no sand was to be seen before. Innumerable little streams overlap and interlace one with another, exhibiting a sort of hybrid product, which obeys half way the law of currents, and half way that of vegetation. As it flows it takes the forms of sappy leaves or vines, [...] resembling, as you look down on them, the laciniated lobed and imbricated thalluses of some lichens; or you are reminded of coral, of leopards' paws or birds' feet, of brains or lungs or bowels, and excrements of all kinds. It is a truly *grotesque* vegetation, whose forms and color we see imitated in bronze, a sort of architectural foliage more ancient and typical than acanthus, chiccory, ivy, vine, or any vegetable

> leaves; destined perhaps, under some circumstances, to become a puzzle to future geologists.[14]

One can first notice that Thoreau emphasizes observation, instead of measuring and cutting, and does not resort to any tools in this sequence, as distinct from mapmaking and the harvesting of ice. His only tool is language, whose sole function here is not to impose order, but to convey the uniqueness of experience. The passage is a striking ode to matter in its sensible materiality and concreteness – fully acknowledging it. One should also note that Thoreau does not offer a gentle, sentimental depiction of spring (no softly twittering birds…), but spring in its raw – or wild – character. He also foregrounds the idea of intricacy ('mixed', 'overlap', 'interlace', 'imbricated') and excess ('bursting out', 'overflowing'), as distinct from linearity and containment (the neat delineation of the map and delimitation of the ice). In a particularly telling way, this passage offers a much more open structure than either the map or the description of the ice harvest: it is indeed impossible to illustrate what is described, i.e., to turn it into an image. Hence the emphasis on the word 'grotesque' (emphasis in the original), which is meant to highlight the idea that this perception does not bow to the prevailing aesthetics of the 'picturesque' (a notion which literally refers to what can be turned into a picture): as he is about to round off the book, Thoreau strikingly chooses to celebrate *imagining* instead of imaging.

In an equally decisive way, he also shifted the perception of the physical world toward an acknowledgment of its autonomy:

> I feel as if I were nearer to the vitals of the globe, for this sandy overflow is something such a foliaceous mass as the vitals of the animal body. You find thus in the very sands an anticipation of the vegetable leaf. No wonder that the earth expresses itself outwardly in leaves, it so labors with the idea inwardly. The atoms have already learned this law, and are pregnant by it. The overhanging leaf sees here its prototype. *Internally* whether in the globe or animal body, it is a moist thick *lobe*, a word especially applicable to the liver and lungs and the *leaves* of fat, (*λείβω*, *labor*, *lapsus*, to flow or slip downward, a lapsing; *λοβος*,

14 Thoreau, *Walden*, pp. 304–5.

> *globus*, lobe, globe, also lap, flap, and many other words,) *externally* a dry thin *leaf*, even as the *f* and *v* are a pressed and dried *b*. The radicals of lobe are *lb*, the soft mass of the *b* (single lobed, or B, double lobed,) with a liquid *l* behind it pressing it forward. In globe, *glb*, the guttural *g* adds to the meaning the capacity of the throat. The feathers and wings of birds are still drier and thinner leaves. Thus, also, you pass from the lumpish grub in the earth to the airy and fluttering butterfly. The very globe continually transcends and translates itself, and becomes winged in its orbit.[15]

As distinct from the passages devoted to the drawing of the map and the harvesting of the ice, this third sequence clearly focuses not on business but on idleness. Thoreau claims the right not to be serious in an ordinary (social) sense, instead emphasizing playfulness and lived experience – although one that is necessarily mediated by language. He also enhances the untranslatable character of the real, literally staging the frantic striving of words (English, Greek, Latin) for an inherently elusive, ungraspable reality, and eventually states the idea that nature 'translates itself' – rather than being processed through human intervention. No verbal or visual translation is possible – language can only gesture toward the world, not rule or tame it: here mankind is no longer master of matter vis-à-vis submissive or compliant nature, but mere witness to a reality that is autonomous, if not rebellious. Thoreau thus resists the desire for domination of physical nature, and accepts the idea of losing control of the real. He voices his commitment to nature's inherent value, as a stark rebuke of the material and financial benefit derived from nature, which is a mere reflection of self-interest. While this passage does not revert to the belief in the bottomlessness of the pond, it nevertheless restores an aura which technology had destroyed or dissolved. Thoreau makes it clear that this is achieved via a commitment to observation – but one that is thoughtful and complexified – as against the mere management of nature. And he resolutely ventures beyond the assertion of order, which mapmaking and the logging of the ice had brought forward. Whereas both science and economic exploitation are meant to produce and shore up the stability of the social order,

15 Thoreau, *Walden*, pp. 306–7.

instability here prevails: in exact proportion to his celebration of the dazzling flux of matter, Thoreau turns his back on the values and privileges of settler colonialism, instead celebrating the *unsettling* power of emancipated perception.

Thoreau subsequently builds on this dynamics to argue for wilderness preservation: the idea of the untranslatable character of the real serves as the rationale for the notion of 'wildness' and the ensuing idea of nature preservation which concludes the chapter:

> Our village life would stagnate if it were not for the unexplored forests and meadows which surround it. We need the tonic of wildness, – to wade sometimes in marshes where the bittern and the meadow hen lurk, and hear the booming of the snipe; to smell the whispering sedge […]. At the same time that we are earnest to explore and learn all things, we require that all things be mysterious and unexplorable, that land and sea be infinitely wild, unsurveyed and unfathomed by us because unfathomable. We can never have enough of Nature. We must be refreshed by the sight of inexhaustible vigor, vast and Titanic features, the sea-coast with its wrecks, the wilderness with its living and its decaying trees, the thunder cloud, and the rain which lasts three weeks and produces freshets. We need to witness our own limits transgressed, and some life pasturing freely where we never wander.[16]

Carefully framing his narrative so as to question standard ways of appropriating the world (scientific inquiry and economic exploitation of natural resources), and propounding a poetics in which the dynamics of writing (trans)figures the processes of nature, Thoreau grounded his environmental advocacy in a broader regard for the indomitable power and enduring secrecy of the physical world. Accepting the existence of nature as exceeding human control and signifying processes shifts the human/nonhuman divide, thus laying the ground for a reshaped humanism. Rather than segregating nature from the human mind, he paradigmatically articulates the wealth of earthly experience, the dense fabric of the world, both material and semantic, in other words the human-nature entanglement. As distinct from a worldview widely prevalent in new materialist approaches, however, entanglement does not mean merging: Thoreau's stance

16 Thoreau, *Walden*, pp. 317–8.

is not a form of posthumanism, in which the human is dissolved within multi-layered networks of material and linguistic constructs, but a redefined humanism, in which the human is understood as more fully realizing itself as a result of its acknowledgment of its interdependency with (and within) the physical world.[17]

The process by which Thoreau, eschewing intellectual privilege as much as economic ownership, gives nature free rein, constitutes a far-reaching thought experiment, redefining the very coordinates of existence. The physical world here appears as unmanageable ('unfathomable'), rather than managed: Thoreau's extraordinary flourish may be regarded as a way of returning nature to its pristine state, after it has been desecrated by human control or plunder. In his thought, the pristine state will primarily be relocated as nature's independence from human constructs. This state is one in which the observer literally experiences disorientation, being requested to forsake their linguistic tools, not to mention their scientific gear or their technological equipment. Thoreau certainly did not mean to advocate an illusory return to an original, unmediated nature, but offered as an essential and indispensable experiment the effort to lose sight of oneself in what one sees. To that extent, Thoreau's thinking has just as much to do with an ethics of displacement as with an experience of place, which first-wave ecocriticism too narrowly foregrounded.[18] But it is just as important to point out that Thoreau did not mean to cancel science nor knowledge – and it is actually because he had moved this far that he was able to read and welcome Darwin's *Origin of Species* almost immediately upon its publication

17 New materialism has become one of the most influential theoretical paradigms in literary studies, and so widespread – and diverse – that offering even a brief overview is beyond the scope of this essay. Notable theorists include Bruno Latour and Jane Bennett. See Bruno Latour, *Reassembling the Social: An Introduction to Actor-Network-Theory* (New York: Oxford University Press, 2007); Jane Bennett, *Vibrant Matter: A Political Ecology of Things* (Durham, NC: Duke University Press, 2010).

18 For an insightful discussion of the different 'waves' of ecocriticism, see Lawrence Buell, *The Future of Environmental Criticism: Environmental Crisis and Literary Imagination* (Oxford: Blackwell, 2005).

late in 1859.[19] Any suspension of knowing is merely methodological, as part of a counterproject seeking to make knowledge a tool for perceiving better, rather than for overcoming estrangement. Nor did he really mean to shun any form of transformation of nature by human labor. Thoreau does not propound a dissolving of the distinction between human and nonhuman, nor a negation of human freedom, as the posthuman challenge would have it, but a more complex view of what being human means, which eventually aknowledges its entanglement with the non-human – and thus constitutes a more balanced – less 'imperial' – form of humanism.

19 For 1859 as a key year in the intellectual history of the nineteenth century, see Anna Pia Ruoppo's essay in this volume.

Andrea Bocchetti*

NIETZSCHE, NATURE AND ITS DOUBLE

Abstract

This article explores Nietzsche's vision of living beings as inherently oriented toward *having*, incorporating, and growing. Through perpetual vulnerability and exposure to otherness, life strengthens and evolves; order and generation arise from disruption rather than equilibrium. Concepts such as *Einverleiben* (incorporation), the impulse to appropriation, and the dual nature of living and inert systems, illuminate the dynamic interplay between self-maintenance and transformation, placing Nietzsche's thought within what we define as *Echontology*, a reading of life through the prism of having, exceeding metaphysics.

Keywords

Nietzsche – Echontology – Incorporation – Double nature – Growth and regeneration

1. *Nature and Life*

The thought of *Einverleibung* – incorporation – is not merely a peripheral moment in Nietzsche's vocabulary. It marks, rather, a locus of originary resonance, where language, life, and force converge.[1]

* Università degli Studi di Napoli Federico II.

1 Unfortunately, it is not possible to provide a critically univocal translation of this term. As will be shown below, the solutions proposed in the various editions of Nietzsche's complete works – while privileging 'assimilation' and 'incorporation' – are diverse and do not appear to establish a clear connection between the term and the underlying concept. See below, § 2.

Not enough has yet been made of its conceptual gravity. What is at stake here is not a question of terminology, but of a hermeneutic of *becoming*: *Einverleibung* articulates, in its very gesture, the tension that holds together Nietzsche's thinking of 'life *as* will to power'.[2]

To read *Einverleibung* is not to enumerate its appearances, nor to classify its uses, but to dwell in the force-field it opens within Nietzsche's corpus. It is less a matter of content than of configuration – less of what is said, than of how saying shapes what remains unsaid. That subtle interplay – always at work in Nietzsche – between the implicit and the creased, the shadowed and the luminous, the latent and the manifest, proliferates once again in the interstices of numerous passages where this notion stirs. In those fissures, something becomes thinkable that demands a certain kind of attentiveness, a tarrying that resists the rush toward conclusion.

What this essay seeks to show is that *Einverleibung* is not only a figure of the will to power, but one of its most effective expressions. It offers a gateway into Nietzsche's physiology,[3] wherein life is no longer grasped as a given, but as a continuous act of assimilation, appropriation, transformation. Through *Einverleibung*, a grammar

---

2 Friedrich Nietzsche, *NF* 7 [45] 1886, KSA 12, 309 (my translation). All quotations from Nietzsche's published works are taken from Cambridge University Press, *Cambridge Texts in the History of Philosophy* series, using the standard translations and editorial attributions. Since no complete critical edition in English yet exists for Nietzsche's posthumous fragments (*Nachgelassene Fragmente*), I provide my own translations for these texts. In all cases, references are given to the *Kritische Studienausgabe* (KSA), ed. by Giorgio Colli and Mazzino Montinari, 15 vols (Berlin/New York: de Gruyter, 1980–2001), indicating fragment number, volume, and page(s). From now on, citations of the posthumous fragments will be marked with the abbreviation NF, followed by the fragment number and KSA reference.

3 It is noteworthy how, in the late Nietzsche's ambitions, the project of a *physiology* of the will to power is programmatically linked to a *morphology* of the will to power as a theory of the affects. On this matter, see the implications drawn in § 2 concerning the central role acquired by the concept of *form* in relation to that of the *body*, as well as the related considerations on the connection with Goethe's thought. See also, in this regard, footnote 20. Nietzsche, *NF* 13 [2-3] 1888, KSA 13, 214-5.

of vitality reveals itself – one that binds together a genealogy of knowledge and a topology of life's relation to world.

To begin grasping the weight of this term, we turn to a decisive distinction Nietzsche offers in aphorism 109 of *The Gay Science*,[4] where he writes:

> Let us beware of thinking that the world is a living being. Where would it stretch? What would it feed on? How could it grow and procreate? After all, we know roughly what the organic is; are we then supposed to reinterpret what is inexpressibly derivative, late, rare, accidental, which we perceive only on the crust of the earth, as something essential, common, and eternal, as those people do who call the universe an organism? This nauseates me. Let us beware even of believing that the universe is a machine; it is certainly not constructed to one end, and the word 'machine' pays it far too high an honour. Let us beware of assuming in general and everywhere anything as elegant as the cyclical movements of our neighbouring stars [...]. Let us beware of attributing to it heartlessness or unreason or their opposites: it is neither perfect, nor beautiful, nor noble, nor does it want to become any of these things; in no way does it strive to imitate man! In no way do our aesthetic and moral judgements apply to it! It also has no drive to self-preservation or any other drives; nor does it observe any laws. Let us beware of saying that there are laws in nature. There are only necessities: there is no one who commands, no one who obeys, no one who transgresses.

In aphorism 109, Nietzsche sets into motion a series of characterizations that articulate a distinction between *life* and *world* – a distinction not merely of entities, but of *natures*, of differing trajectories of emergence. Though the organic, as Nietzsche notes, arises as a derivation – rare, contingent, a deviation from the inanimate – it nevertheless proceeds according to a typological composition that diverges, at least partially and in appearance, from the chaotic tendencies of the inorganic universe.

What Nietzsche underscores here is not a metaphysical rupture but a differential in rhythm, in articulation. Life does not negate the drift

---

4 Nietzsche, *The Gay Science*, 109, ed. by Bernard Williams, trans. by. Josefine Nauckhoff (Cambridge: Cambridge University Press, 2001), pp. 109–10; KSA 3, 467–8.

toward disorder; it interrupts and slows it, folding it into a specific configuration – temporary, forceful, expressive. The so-called normative order of the world – its 'beauty', its 'laws', its 'ends' – does not precede life; on the contrary, it is drawn from it, imposed by it, extracted through the metaphoric gestures of a being who has needed to make sense of his place within it: simplify to dominate. Thus, what has been received as 'spiritual' is nothing but an intensification of the organic, where it resounds something more intimate, more ancient, more originary; the organic world itself – not as static substance, but as a field of forces through which life configures itself *spiritually*: interpreting, organizing, projecting, and propagating within the impulse to grow and dominate:

> The commanding element (whatever it is) that is generally called 'spirit' wants to dominate itself and its surroundings, and to feel its domination: it wills simplicity out of multiplicity, it is a binding, subduing, domineering, and truly masterful will. Its needs and abilities are the same ones that physiologists have established for everything that lives, grows, and propagates.[5]

Spirit, for Nietzsche, is not a transcendent principle imposed upon life, but the cumulative expression of life's form and rhythm – its way of happening. To probe the constellation of forces that constitute the so-called cognitive apparatus, or the universe of things, or the ontologizing tendencies that shape philosophical and scientific thought in their attempt to frame both world and human within clearly defined categories, it is not sufficient to verify the analogical fit between thought and being, nor to delimit the transcendental space of possibility. Rather, as Nietzsche insists, one must draw from those knowledges that allow the tangled lineages connecting inorganic matter, organic processes, and spiritual life to become discernible: namely, physiology, biology, botany, zoology.

Nietzsche's encounter with these disciplines enables a rewriting of the phenomenology of the spiritual – one that locates in organic

---

5 Nietzsche, *Beyond Good and Evil: Prelude to a Philosophy of the Future*, 230, ed. by Rolf-Peter Horstmann and Judith Norman, trans. by Judith Norman (Cambridge: Cambridge University Press, 2002), p. 121; KSA 5, 167.

dynamics, even in the behavioural mechanisms of the simplest cells, the traces of what we later come to name as 'the spiritual'. Thinkers such as Goethe, Darwin, Roux, Virchow, Haeckel, and Nägeli[6] contribute to this subterranean *genealogy of the human*,[7] which does not merely expose the arbitrariness or limitations of symbolic apparatuses but reconnects them to the grammar of the living, without thereby succumbing to reductive determinism. That there are levels of continuity between biochemical and metabolic rhythms and the realm of spiritual life does not imply a collapse of one into the other, but rather a species of differential oscillation. Nietzsche holds fast to the chaotic texture of the world – and it is precisely into chaos that regularities must be reintegrated. As he himself puts it, chaos is not 'not in the sense of a lack of necessity but of a lack of order'.[8]

The global flow in which life is placed and entangled gives rise to incessant alterations and proliferations. What appears stabilized is continuously subjected to forces of disruption and reconfiguration. Life is no exception. On the contrary, its very unfolding or extinction depends on this tragic relation to its *Außenwelt* – to the external world – and more precisely, on its capacity to incorporate it (*Einverleiben der Außenwelt*).[9] It is in this sense that the living *resists*,[10] and likewise, in this very sense, that it complexifies through growth and metamorphosis.

---

6 For a comparison between Nietzsche and scientific thought, see: Babette Babich, *Nietzsche's Philosophy of Science. Reflecting Science on the Ground of Art and Life* (New York: State University of New York Press, 1994); Francesco Moiso, *Nietzsche e le scienze* (Turin: Rosenberg & Sellier, 2020); Barbara Stiegler, *Nietzsche et la vie: Une nouvelle histoire de la philosophie* (Paris: Gallimard, 2021).

7 We borrow the expression from the work in honor of Aldo Masullo: *Genealogia dell'umano. Saggi in onore di Aldo Masullo*, ed. by Giuseppe Cantillo and Felice Ciro Papparo (Naples: Guida, 2000).

8 Cf. *supra*, note 4.

9 Nietzsche, *NF* 2 [92] 1885, KSA 12, 106–7.

10 'A multiplicity of forces, connected by a common process of nourishment (*Ernährungs-Vorgang*), is what we call "life". To this process of nourishment – as a means of its enablement – belongs everything so-called feeling, representing, thinking, that is: (1) a resistance (*Widerstreben*) against all other forces; (2) a shaping of them into forms and rhythms; (3)

Nietzsche does not posit a flat correspondence between cellular processes and the symbolic-spiritual domain. What he sees instead is a movement of fracture, restructuring, and transformation – where a basic homology gives way to a proliferating field of expressive forms. *Einverleibung* becomes crucial precisely because it marks a distance: a gap between the organic homologies and the danger of relapsing into a metaphysics of essences. What defines the living and the spiritual for Nietzsche is not the metaphysical tension of potential being – but the potential tension of life itself, which *tends* to incorporate, which interprets the organic by *imposing upon it the traits of being*.

Nietzsche's task is not to delineate a symbolic order merely distinct from the inorganic natural order, nor to identify a simple internal homology between life and spirit, but to show how the symbolic order is the complex precipitate of a state of the body ('There are *only* bodily states [*leibliche Zustände*]: the spiritual ones are consequences and symbolism').[11]

An impression, an error, arising from life's self-representation – or more precisely, from the necessary illusion that compels the living to stabilize, to unify, to identify, by projecting outward what has first imposed itself as an inward affect.

That *tensional line*, which allows glimpses of regularity, stabilization, and repetition in the living being's apprehension and self-apprehension, extends first into the epistemic space of law, and then into the symbolic domain of beauty, order, and purpose – pervading the entire universe of the humanized and spiritualized world, metabolized and incorporated. It is no accident, then, that Nietzsche likens human knowledge to a kind of stomach.[12]

It becomes clear how, for Nietzsche, even mechanistic explanations are epistemological prejudices – sustained and derived from biological and physiological processes. To deconstruct the claims of knowledge,

---

an evaluation in terms of incorporation or rejection (*auf Einverleibung oder Abscheidung*)'. Nietzsche, *NF* 24 [14] 1883, KSA 10, 650–1.

11 Nietzsche, *NF* 9 [41] 1883, KSA 10, 358.

12 'Human knowledge has become refined, and in the end so proud of its refinement that it no longer wants to hear that it is not an end, but a means – or even a tool of the stomach – if not itself a kind of stomach (*eine Art von Magen*)!'. Nietzsche, *NF* 38 [10] 1885, KSA 11, 609.

one must infiltrate the *potential dynamics* of living beings: their movements, articulations, organizations, tendencies, impulses, and their relations with the world – or rather, with their own world.[13]

Error, therefore, is not read by Nietzsche as a cognitive defect ('*Truth* does not designate an opposition to error, but rather the position of certain errors in relation to other errors – for example, that they are older, more deeply incorporated [*tiefer einverleibt*], that we do not know how to live without them'[14]), but rather as a necessary effect of the expansive character of the living ('To err is the condition of life, and indeed to err profoundly [*Grunde Irren*]'[15]). The world is not given, then, but created ('Thus arises our world, our entire world: and to this whole world, belonging solely to us and created only by us, there corresponds no supposed 'true and proper reality [*eigentliche Wirklichkeit*]'[16]) – that is, the stabilized resultant of an internal configuration perceived as enduring.

The living, in the direction of creating its own world, *imposes* upon the flow that envelops it the categories drawn from the first instance that allows it to say: the *I*. Yet this creation is not only morphic elaboration; it is the direct consequence of a prolongation inherent in biological processuality – that is, growth and nourishment. Living beings create their world by appropriating it, or better, by *incorporating* it.[17]

---

13 Cf. Jakob von Uexküll, *A Foray Into the Worlds of Animals and Humans: With a Theory of Meaning*, trans. by Joseph D. O'Neil (Minneapolis/ London: University of Minnesota Press, 2010); Edgar Morin, *L'écologie généralisée* [*Oikos*], in Id., *La Méthode 2. La vie de la vie* (Paris: Seuil, 1980); Gilles Deleuze, *Le Pli. Leibniz et le baroque* (Paris: Les Éditions de Minuit, 1988).

14 Nietzsche, *NF* 34 [247] 1885, KSA 11, 503.

15 Nietzsche, *NF* 11 [162] 1881, KSA 9, 504.

16 Nietzsche, *NF* 38 [10] 1885, KSA 11, 609.

17 Speaking abstractly of 'world' in a homological and reductively unitary image is at present contrary to any scientific assessment. More accurately, one should speak of ecosystems, that is, complex relationships between the biotope and the biocenosis of a given environment. All this obviously makes any attempt to quantify the number of 'worlds', as well as their relationships, difficult, and even whether these can be considered 'systemic'. On this subject, see the valuable work of Nicola Russo, *Filosofia e ecologia* (Naples: Guida, 2000).

It is precisely due to this incorporation that the assimilation of the world can never be integral or global: living beings absorb portions of the world, incorporating the external ('Life is not the adaptation of internal conditions to external ones [*Anpassung innerer Bedingungen an äußere*], but Will to Power – which, from within, increasingly subjugates and incorporates [*sich unterwirft und einverleibt*] the 'external' into itself'[18]), growing and metabolizing according to the limit of corporeality. And although the appropriative feeling knows no limit, what a body appropriates cannot exceed its biochemical assimilative capacities.

Thus, the incorporative action is, on the one hand, elaborative: the flow of the (own) world must – for it to be absorbed – be reduced, simplified, digested. In other words, the living must render it bearable.

On the other hand, incorporation follows a selective logic: the exchange that the living establishes with the world (to the extent that Nietzsche speaks of overcoming the distinction between internal and external world[19]) is never total, but in relation to nourishing, metabolic, and reproductive instances. The functional organization of parts in every living being is thus the direct consequence of its incorporative apparatus, which becomes the measure of its formation and metabolic capacities (an ecometabolism, we might say, alongside biochemical metabolism).[20]

18 Nietzsche, *NF* 7 [9] 1886, KSA 12, 295.

19 Cf., e.g., Nietzsche, *NF* 9 [41] 1883, KSA 10, 358: 'To separate internal and external world (*Außere und innere Welt*) as the metaphysicians do is already a judgment of the senses (*Sinnen-Urtheil*)'.

20 Paola Giacomoni provides a clear and effective exposition of Goethe's idea concerning the meaning of *form*, an idea that seems to find resonance within Nietzsche's physiological perspective: 'Undoubtedly, the internal-external relation is also traced in these analyses of animal skulls: the search for a character, or at least a particular instinctual disposition in the lines of a skull, is quite evident. Clearly, this is not a purely formal investigation, but rather the identification of the meaning of a *form*, of its adequacy to express a certain specific disposition. Yet such disposition, even if at times expressed in terms of "character", is above all understood as a functional organization – as a set of physical aptitudes that can be explained by the animal's specific environment, by its way of feeding, digesting, and moving'. P. Giacomoni, *Le forme e il vivente. Morfologia*

2. Einveleibung *as a Concept.*

Before delving into the depths of *Einverleibung*, it is useful to offer some orientation regarding its chronological geography, including its diversified inflections,[21] as well as the terminological spectrum that accompanies it and composes Nietzsche's phenomenology of incorporation. The term *Einverleibung* appears diffusely, yet with a marked insistence beginning in 1881[22] (the year in which it occurs most abundantly), re-emerging with notable persistence in the following years, peaking particularly in the period 1884–1885. *Einverleibung* receives, in any case, much greater space in the materials that make

---

e filosofia della natura in J. W. Goethe (Naples: Guida, 1993), pp. 43–4; my translation). Nietzsche was undoubtedly influenced by Goethe's scientific writings, employing in various forms the key concepts of *form*, *metamorphosis*, and *morphology* – to the point of defining psychology as a 'morphology of the will to power'. The Goethe–Nietzsche relation would in fact warrant a broader investigation, particularly regarding the influence of Goethe's science on Nietzsche's thought and the terminological borrowings employed by the latter.

21 Below are the most relevant passages, for the purposes of our discussion, in which *Einverleibung* and its inflected forms occur: *Human, All Too Human*, II, 31; *Daybreak*, 281, 285; *NF* 11 [141]; *NF* 1881; *NF* 11 [162] 1881; *NF* 11 [197] 1881; *NF* 11 [268] 1881; *NF* 11 [2] 1881; *NF* 11 [164] 1881; *NF* 11 [193] 1881; *NF* 11 [262] 1881; *NF* 11 [314] 1881; *NF* 11[335] 1881; 11 [316] 1881; *NF* 11 [320] 1881; *NF* 11 [309] 1881; *NF* 11 [302] 1881; *NF* 11 [335] 1881; *NF* 11 [134] 1881; *NF* 12 [40] 1881; *NF* 14 [13] 1881; *The Gay Science*, 11, 83, 110; *NF* 12 [39] 1883; *NF* 7 [107] 1883; *NF* 7 [173] 1883; *NF* 20 [10] 1883; *NF* 24 [14] 1883; *NF* 26 [448] 1884; *NF* 25 [155] 1884; *NF* 25 [409] 1884; *NF* 25 [437] 1884; *NF* 25 [8] 1884; *NF* 34 [247] 1885; *NF* 36 [22] 1885; *NF* 38 [10] 1885; *NF* 40 [4] 1885; *NF* 40 [34] 1885; *NF* 2 [92] 1885; *NF* 5 [64] 1886; *NF* 5 [65] 1886; *NF* 5 [82] 1886; *NF* 7 [53] 1886; *NF* 7 [2] 1886; *NF* 7 [9] 1886; *Beyond Good and Evil*, 230, 259; *On the Genealogy of Morality*, 2, 1; *NF* 9 [151] 1887; *NF* 10 [18] 1887; *NF* 11 [71] 1887; *NF* 11 [111] 1887; *NF* 14[174] 1888.

22 The material produced during this year constitutes, as is well known, the backbone of *Daybreak* and *The Gay Science*. In these works, however, the term *Einverleibung* does not occur with great frequency. Yet, although its use is not abundant, its semantic placement can legitimately be understood as a compensatory thrust counterbalancing its exoteric rarity.

up the *Nachlass* (Posthumous Fragments) rather than in the published works. This may have contributed to a more peripheral consideration of its relevance, but – as in other cases – it is not altogether unfounded to regard the *Nachlass* as a crucial space wherein the intensity of a concept attempts to break forth.

Einverleibung evidently refers to incorporation (according to the morphological composition given by the prefix *ein-* and the noun *Leib*). From the perspective of translations of this term, however, probably due to a not always recognized conceptual weight, the terms used are varied and often lacking in congruence. In the Italian edition, curated by Colli-Montinari, although '*incorporazione*' is the most literal translation, other solutions may also be found, such as *assimilazione* (which is preferred more often than 'incorporazione'), *incarnazione*, or *prendere corpo*. The same holds for the French, English, and Spanish editions. French editions use the term *assimilation*, but also *incorporation*; Spanish editions oscillate between *incorporación* and *asimilación*; and in English, one encounters *embodiment*, *incorporation*, and *assimilation*. The difficulty, however, does not lie in the more or less correct choice of translation. Rather, it lies in the widespread and composite use of expressions related to *Einverleibung* (and its inflections).

Generally speaking, in translations there is a predominance of interchangeability between *incorporation* and *assimilation*. The latter solution is by no means inappropriate, primarily because Nietzsche himself speaks, in relation to *Einverleibung*, of an activity that produces equality – or more precisely, a *making similar* (*Anähnlichung)*,[23] a *making equal* (Gleichsetzung),[24] an *equating*;

23 Cf., e.g., Nietzsche, *Beyond Good and Evil*, 202, p. 90; KSA 5, 124. The translator of the Cambridge edition renders the term *Anähnlichung* as 'approximation', a choice that does not fully capture its semantic nuance. Derived from *ähnlich* ('similar') with the prefix *an-* indicating a movement toward similarity, *Anähnlichung* refers not to an imprecise approach (*Annäherung*), but rather to a process of assimilation or *becoming-similar*. As was observed for the term *Einverleibung*, this choice to favor terminology that prioritizes clarity over conceptual precision further testifies to the fact that this conceptual field has not yet been fully recognized for its specificity within Nietzschean thought.

24 *NF* 2 [92] 1885, KSA 12, 107.

but he also says: '*Assimilation (das Assimiliren) is to make [...] equal to oneself*'.[25] Nietzsche, therefore, is not unfamiliar with the specific term of *Assimilation*[26] and often employs it alongside *Einverleibung*.[27]

It is consequently necessary to ask whether Nietzsche considers the two concepts as synonymous; whether he refers to two distinct 'moments' within the same expansive-digestive dynamic of the living (either consequential, or with one being the general aspect of the other), or whether they are two synecdochic expressions of the will to power.

What we shall endeavour to demonstrate is that *Einverleibung* holds an autonomous signification within Nietzschean thought and that, although often accompanied by additional elements pointing to the expansive character of life, it encompasses a peculiarly general aspect capable of defining, in a broader sense, the principle upon which his phenomenology is constructed – of which the term in question is but a manifestation. In other words, we aim to elucidate the terminological hierarchy to which Nietzsche alludes, the degrees of internal subordination within the same issue: the characteristic constellation of the will to power, whose dynamic transcends *being* toward the pre-dominance of *having*.

---

25 *NF* 11 [134] 1881, KSA 9, 491.

26 Cf., e.g., *NF* 11[134] 1881; *NF* 11 [182] 1881; *NF* 11 [268] 1881; *NF* 11 [270] 1881; *NF* 11 [315] 1881; *NF* 7 [33] 1883; *NF* 7 [86] 1883; *NF* 7 [95] 1883; *NF* 25 [179] 1884; *NF* 25 [356] 1884; *NF* 34 [55] 1885; *NF* 34 [252] 1885; *NF* 47 [2] 1885; *Beyond Good and Evil*, 36.

27 Cf., e.g., *NF* 11 [134] 1881, KSA 9, 491: 'Wenn wir die Eigenschaften des niedersten belebten Wesens in unsere "Vernunft" übersetzen, so werden moralische Triebe daraus. Ein solches Wesen *assimilirt* sich das Nächste, verwandelt es in sein Eigenthum (Eigenthum ist zuerst Nahrung und Aufspeicherung von Nahrung), es sucht möglichst viel sich *einzuverleiben*, nicht nur den Verlust zu compensiren – es ist habsüchtig. So wächst es allein und endlich wird es so reproduktiv – es theilt sich in 2 Wesen. Dem unbegrenzten Aneignungstriebe folgt Wachsthum und Generation. – Dieser Trieb bringt es in die Ausnützung des Schwächeren, und in Wettstreit mit ähnlich Starken, er kämpft d.h. er haßt, fürchtet, verstellt sich. Schon das *Assimiliren* ist: etwas Fremdes sich gleich machen, tyrannisiren – Grausamkeit' (*italics mine*).

A first step toward delineating the relationship between the will to power and *Einverleibung* can be taken in the direction of what Nietzsche designates as the *Haben-wollen*:[28]

> One must want to have more than one has (*mehr haben wollen als man hat*), in order to become more (*um mehr zu werden*). For thus sounds the doctrine that is preached to everything that lives, by life itself: the morality of development (*Entwicklung*). Having and wanting to have more (*Haben und mehr haben wollen*) – growth (*Wachsthum*), in a word – that is life itself (*Leben selber*).[29]

Life itself is *having-will*, that is, growth – namely, a constant tension toward appropriation, unfolding through the entire prism of spiritual apprehensions. The living is driven by an impulse that lacks both precise direction and limit; from it follows growth and regeneration as an extension of the expansive impulse: 'Growth and generation (*Wachsthum und Generation*) follow from the unlimited impulse of appropriation (*Aneignungstriebe*)'.[30] The term employed here by Nietzsche, *Aneignung* (appropriation), represents one of the pivotal nodes of *Einverleibung*: the two terms frequently appear side by side, where the former stands as a reverberation of the assimilative-incorporative movement, as well as the principle from which nourishment follows: '*Nourishment (*Ernährung*) [is] merely a consequence of the insatiable appropriation (*unersättlichen Aneignung*), of the will to power*'.[31] Life thrusts itself outward according to an apprehensive dynamic, absorbing through the body that which it seeks to appropriate:

> Life itself is essentially appropriation (*Aneignung*) […] of the alien […], imposition of one's own form, incorporation (*Einverleibung*) […].

---

28 Nietzsche employs this expression both in its uninflected form, as in the passage cited below, and in its substantivized form, as in fragment *NF* 11 [19] 1881, KSA 9, 449: 'Perhaps all moral drives can be traced back to the *having*-will and the *holding*-will (das *Haben*-Wollen und *Halten*-wollen)'.

29 Nietzsche, *NF* 37 [11] 1885, KSA 11, 586.

30 Nietzsche, *NF* 11 [134] 1881, KSA 9, 491.

31 Nietzsche, *NF* 2 [76] 1885, KSA 12, 96.

> Even a body [...], if it is living and not dying, will have to embody the will to power and seek to grow, spread, grasp, and gain dominance.[32]

Here begins to emerge the crucial locus upon which the entire appropriative process condenses: the body. For appropriation to occur, life must *incorporate itself by incorporating*: the expansion of the will to power takes form through a *becoming-body*. Or rather, the body potentiates itself by growing – that is, by incorporating what it assimilates and appropriates. The evolutionary alteration of the living being is therefore not unidirectional: it absorbs and incorporates what it assimilates, but what it assimilates simultaneously constitutes and alters it; its integrity, its *sameness*, is continuously compromised.[33]

It must also be emphasized that the appropriative-incorporative impulse, while reiterating the peculiarity of the organic tendency, does not develop into monotonous or crystallized forms. Its recurrence is spiralized, displaced into more articulated and complex forms, within a circle that expands toward the indeterminate and the complex: the organic diversifies by organizing itself (as Morin states, '*la diversité est l'ingrédient et le produit de toute organisation vivante*'[34]). It is therefore neither a systemic nor predetermined nor predictable movement; yet it specifies itself in form. *Having*, therefore, does not cease to incorporate itself into new forms, precisely because its mode of apprehension does not follow the logic of fullness (albeit such a '*dream*' remains the horizon upon which the imaginary enjoyment of today is structured); quite the opposite:

> The concept of having (*Der Begriff des Habens*) continually refines itself; we understand more and more how difficult it is to have, and how apparent possessions still manage to elude us – thus we drive having

---

32 Nietzsche, *Beyond Good and Evil*, 259, p. 153; KSA 5, 207.

33 Hans Jonas, speaking of the sameness of life, says: 'Sameness, while it lasts (and it does not last inertially, in the manner of static identity or of on-moving continuity), is perpetual self-renewal through process, borne on the shift of otherness. This active self-integration of life alone gives substance to the term "individual"'. Hans Jonas, *The Phenomenon of Life. Toward a Philosophical Biology* (Evanston, Illinois: Northwestern University Press, 1966, p. 79).

34 Morin, *La Méthode 2. La vie de la vie*, p. 305.

> into the finer: until finally complete apprehension of the thing becomes the precondition for striving for it; often complete apprehension already suffices as possession, it has no hiding place left and can no longer escape us […]. Earlier stages, for example: imagining a thing and then believing that one fully possesses it (*es ganz besitze*) […]. We conceive the things we can attain so that their possession appears extremely valuable to us […]. We first make a rough calculation of what we can actually seize – and now our imagination is active in making these future possessions extremely valuable to us […]. We seek the philosophy that suits our possession, i.e., gilds it.[35]

Having, then, does not unfold as a mere thrust that, in its most refined forms, culminates in possession; its recurrence arises precisely because of that which escapes, that which demands the new, and which does not stabilize in what is possessed, except in the imaginary instance of fullness. Strictly speaking, *having* always repeats a missed encounter: nothing is ever fully possessed. What *having*, on the contrary, reveals is that life establishes interactions, relations, exchanges, flows, which cannot be reduced either to encounters between essences or to apparent microeconomic bonds. Nietzsche, in this regard, seems to seek a rehabilitation of *having*, despite the trivial denigration that has become embodied in its common usage. In this sense, his discourse resonates with another theme, equally relegated to the realm of the deplorable, *egoism*:

> Extend the concept of nourishment (*Ernährung*); do not mismanage your life as those do who merely have their preservation (*Erhaltung*) in view. We must not let our life slip through our hands by a 'goal' – but rather harvest the fruits of all seasons for ourselves. We want to seek after others, after everything outside us, as after our nourishment (*unserer Nahrung*). Often it is the fruits that have just ripened for our year. – Must one always have only the egoism of the robber or thief? Why not that of the gardener? Joy in tending others (*Freude an der Pflege der Andern*), as in a garden![36]

This passage brings to light the expansive bearing of the organic's inherent tension, refined into a form of nourishment that does

35 Nietzsche, *NF* 11 [19] 1881, KSA 9, 449.
36 Nietzsche, *NF* 11 [2] 1881, KSA 9, 441.

not merely aim at plundering the other or the foreign. Rather, the image of 'tending' (*Pflege*) gestures toward a nourishing modality that signifies a kind of *care*, one that absorbs and assimilates insofar as it potentiates, not merely conserves itself: 'Conquering (*Erobern*) – is the natural consequence of an excess of power: it is the same as creating and engendering (*Schaffen und Zeugen*), that is, incorporating (*Einverleiben*) one's own image into alien matter'.[37]

There is thus a excessive, dispersive, dissipative matrix inherent in potentiation; it is not a 'calculation' in a utilitarian sense. Appropriation does not proceed from an accumulative purpose that simply collects what it acquires: rather, it is the necessary outward diffusion of an energy that overflows.

Self-preservation, then, is the weakened form of potentiation, that which intends to endure as duration rather than as overcoming.

Egoism, therefore, moves in a counter-movement, thrusting itself toward the other on the ridge of its own expansion: from this arises nourishment as a process bearing the mark of the organic, yet in its richest, most 'gilded' formula – that is, its most intimate form, its most *intimate need* – which does not merely possess, nor possess fully:

> *Possession possesses* (Der Besitz besitzt). It is only up to a certain degree that possession makes humanity more independent and freer; one step further – and possession becomes master, the possessor a slave: as which the latter must sacrifice to the former his time and his reflection, and henceforth feels compelled to social intercourse, nails himself to a place, and feels incorporated [*einverleibt*] into a state – all perhaps against his inner and essential need.[38]

Possession, then, can derive its 'grammar' from the organic only insofar as it reaches a certain degree, where the appropriative impulse does not pertain to a thirst for possession, nor aims at possession as

37 Nietzsche, *NF* 7 [107] 1883, KSA 10, 278.

38 Nietzsche, *Human, All Too Human*, II, 317, transl. by R. J. Hollingdale (Cambridge: Cambridge University Press) 1996, p. 284 (minor adjustments made for terminological fidelity to the German text); KSA 2, 507.

an end, but as a means to potentiation: like interpretation, it is a process and not a mere act.[39]

Egoism, like having, is re-proposed by Nietzsche from the perspective of the vital potential dynamic: it is not about reading the living as the centering of the *ego* in relation to the world, but about identifying within the constitution and formation of the *ego* that *Trieb* internal to organic processes which underlie the evolution and expansion of the living. The incorporative impulse, although extending infinitely, halts at the limit of what is *assimilable* and bearable.

Throughout the process of generation[40] the organism is subjected to persistent self-regulation, confronted with ever more complex degrees of incorporation, and requiring correspondingly more articulated elaborations. Having does not entail mere accumulation: the living body is perpetually driven to refine its metabolic functions through the resistance of the other and the limits it can bear. As Barbara Stiegler splendidly observes: the living body's assimilative activity depends precisely on being affected by alterity, on facing resistance and complexity that force it to transform itself.

> La nutrition implique un autre, l'assimilation une excitation, la régulation un dérangement. En ce sens, l'irritabilité du corps vivant, sa capacité à être altéré par une altérité qui lui arrive d'ailleurs et qui le trouble, est la condition même de l'activité assimilatrice et le critérium ultime du vivant. [...] Si le corps vivant assimile (comprend, pense et veut le même), c'est justement parce qu'il est diversement affecté, parce qu'il éprouve du toujours différent qui lui résiste, parce qu'il a

39 Cf. *NF* 2 [148] 1885, KSA 12, 139–40.

40 It should be emphasized here – although this aspect would require extensive treatment – that just as 'generation' coincides with a 'disintegration that arises from the impotence of the dominant cells to organize the assimilated substances', in the same way creation (in its *generating*), as it grows in complexity and proceeds from an excess seeking to 'encounter' the world, is not governed by an impulse toward pleasure but by a *torment*, that is, by the degree of unendurability of what is incorporated: 'Even if, compared to men of simpler epochs, we have the advantage of a taste for rare and unheard-of dishes, we know true digestion, true living, true assimilation, incorporation (*das Einverleiben*), almost only as torment (*Qual*)' (*NF* 25 [8] 1884, KSA 11, 11).

> affaire à une énorme complexité, dont l'unité et l'unification sont à gagner. Tel est le sens complet de la nutrition comme incorporation : l'assimilation bien sûr, mais aussi et surtout l'exposition de nos procédures assimilatrices à l'altérité prodigieuse qui les déborde et qui les contraint à se transformer elles-mêmes.[41]

In other words, its tension toward growth constitutes a constant conflict between the strength of its unity (its form) and its capacity to transform – assimilating the other, thereby being altered – without being deflagrated.

3. *Echontology*

At first glance, it might seem that the concepts of *nourishment* (*Ernährung*), *growth* (*Wachsthum*), *generation* (*Generation*, *Zeugung*), *assimilation* (*Assimilation*), *appropriation* (*Aneignung*), and *incorporation* (*Einverleibung*) are situated on the same plane of relevance. What one can observe, instead, is that Nietzsche appears to assign a privileged position to the last of these: *Einverleibung* More specifically, one must add that incorporation is driven by an *impulse* (*Trieb*). One could, in other words, attempt to synthesize the *will to power* as an *impulse toward incorporation*. One initial step allows us to follow this hypothesis:

> Such a being assimilates (*assimilirt*) what is nearest, transforming it into its own property (property is, at first, nourishment (*Nahrung*) and the storing up of nourishment); it seeks to incorporate (*einzuverleiben*) as much as possible, not merely to *compensate* for loss — it is *covetous*. Only in this way it grows (*wächst*), and finally it becomes so *reproductive* that it divides into two beings. Growth and generation follow from the unlimited impulse of appropriation.[42]

From this passage it can be inferred that (1) through *assimilation*, there is *appropriation* – in which property is identified with nourishment – and that (2) *incorporation* produces *growth*. Finally,

41 B. Stiegler, *Nietzsche et la vie*, p. 52.
42 Nietzsche, *NF* 11 [134] 1881, KSA 9, 491.

that *growth* and *generation* follow from the *appropriative impulse*. Elsewhere, Nietzsche – particularly in the passage already mentioned – confirms that *nourishment* is a derivative effect: *'Nourishment [is] merely a consequence of the insatiable appropriation* (*unersättlichen Aneignung*)*, of the will to power'*.[43]

The constellation of terms thus seems to converge toward a specific point that gathers the others: the *Aneignungstrieb*. The *appropriative impulse*, as *will to power*, coincides with the expansive thrust of life that absorbs, *assimilates* the foreign, that tends toward the other and the outside. Now, what, then, is the relation between *assimilation*, *incorporation*, and *appropriation*? While *appropriation* (*Aneignung*) and *assimilation* (*Assimilation*) are respectively marked as an impulsive tension toward 'making one's own' and as nourishing acquisition – inasmuch as both contain the root *eignen* (to make proper) and the notion of *the similar* – they nonetheless evoke two indeterminate placements (where, indeed, is that which becomes proper located? And where, that which is rendered similar?).

In contrast, *Einverleibung*, precisely because it refers to the *Leib*, the body, enriches both the appropriative impulse and assimilation by centering them within the key site of Nietzschean physiology. Not only because the body is the horizon repeatedly evoked as the methodological starting point, but more importantly because it is only in the body that the tragic and tormented dimension of *appropriation* is defined: the organic metabolism, within which *assimilation* is inscribed, cannot be understood without a body that incarnates what it appropriates.

The *appropriative impulse* cannot be conceived as a mere force that passes through the body – unless one reverts to a metaphysical principle: only the body lends *sense* to that impulse. In *assimilation*, the body *becomes* by incorporating – that is, by selecting portions of the world, organic and inorganic, and embodying them through assimilation, while not incorporating everything it assimilates. As with every metabolic function, the elaboration of nourishment always contains an expelled part:

43 Nietzsche, *NF* 2 [76] 1885, KSA 12, 96.

> Thus do we all live! – we seize things greedily and fix upon them with insatiable eyes; then we extract from them, just as greedily, whatever tastes good to us and is useful [...] to other human beings and to nature, above all everything we have devoured without being able to incorporate it (*einverleiben*): our excrements (*Excremente*).[44]

Thus, if digestion and assimilation leave a remainder, incorporation is the full transcription of the Other onto the body. Between the impulse of *appropriation* and growth stands the body, whose becoming answers to the resolution of a struggle against all that resists: 'What is "passive"? To resist and to react. To be impeded in forward movement: thus, an acting of resistance and reaction. What is "active"? To tend toward potency (*nach Macht ausgreifend*)'.[45] *Having*, resolved in the body, is therefore the 'grammar' of every organic becoming toward *potency*: 'All self-expansion (*Sich-ausbreiten*), all incorporation [*Einverleiben*], all growth (*Wachsen*) is a striving against what resists [...]. What do the trees of a primeval forest struggle with one another for? For "happiness"? – For potency...'.[46]

One assimilates in order to incorporate, one metabolizes (selective resistance) in order to incorporate, one grows in order to incorporate, one *appropriates* the exterior in order to incorporate, one nourishes oneself in order to incorporate: in sum, the living body is the potential field *par excellence* of life's tension toward *having*, and *Einverleibung* the law of its becoming-body.

It seems clear, ultimately, how Nietzsche's thought is oriented toward *having*, and how the *will to power* belongs to the domain of *havings* (*having-willings*). One might perhaps speak in this regard of an *Echontology*.[47]

---

44 Nietzsche, *NF* 14 [13] 1881, KSA 9, 626.

45 Nietzsche, *NF* 5 [64] 1886, KSA 12, 209. Although we adhere to the conventional English rendering 'will to power', we shall, where context allows, employ 'potency' to translate *Macht*, as we believe it better captures the nuanced sense of potentiality with which Nietzsche invests the term.

46 Nietzsche, *NF* 11 [111] 1887, KSA 13, 52.

47 We allow ourselves the use of this neologism that we already employed it elsewhere [cf. Andrea Bocchetti, 'Du Même au pareil. Descartes,

Having, as *Habenwollen*, in this sense, could be the modality through which Nietzsche transcends the dimension of being, distancing himself from any metaphysical interpretation of the *Wille zur Macht* even in visions of its completion. For Nietzsche, *being* appears as an image rigidified, slowed, crystallized of a movement, even when it presumes temporality or historicity. It retains, fatally, only what persists (unity, identity, equality), excluding what resists or escapes movement (in physics, this field is sometimes assigned to the negligible, to what in practice does not merit attention).

Conversely, Nietzschean *Echontology* considers *being* as an illusory abstraction of a slowing: every structure, every form, every territory, however apparently configured through elements in systemic connection, is characterized not so much by those connections, as by the endogenous energy required so that the persistent tendency toward alteration may be overwhelmed (that which he calls *health*, *Gesundheit*). Every component thus possesses the constant impulse to withdraw or to acquire greater space within the structure; every component therefore tends toward disintegration, by reason of its *incorporative impulse*. Not in the sense of a regressive impulse toward annihilation (not the Freudian *thanatos*, for example), but in the sense of an insistent compromise of the threshold that determines its state of equilibrium. Through a careful examination of late nineteenth-century physiological and biological thought, it becomes evident to Nietzsche that the endurance and strengthening of living beings are intimately bound to their perpetual risk of annihilation: only through a certain disposition to alteration, through the refinement of irritability and of

Nietzsche et l'égologie de la substance', *Les études philosophiques*, 184.4 (2018), pp. 495–518], in order to attempt a synthesis of the unprecedented theoretical and terminological field inaugurated by Nietzsche through a new concept. The expression draws on the etymological structure of classical ontology: just as *òntos* (ὄντος) is the genitive singular of the present participle of *eimi*, 'to be' (ὤν, being), and denotes 'of the being', the term *échontos* is analogously the genitive singular of the past participle of *échō*, 'to have', and indicates 'of the one who has / of the having'. Its combination with *lògos* evokes the classical construction of *lògos tou òntos*, now reformulated as *lògos tou échontos* to signify 'the discourse of the having', according to a beyond-metaphysical perspective.

incorporative capacity – and thus only through a certain degree of vulnerability – can living beings regenerate and fortify themselves. It is solely the rupture of their morphological equilibrium that allows them to evolve. In this sense, every component is continuously subjected to a complex of solicitations that nourish it, alter it, test its resilience; meanwhile, that very resilience coincides with the distance that separates each organic aggregate from the mortiferous equilibrium of disintegration (after all, these are nonlinear transitions toward different states of equilibrium). Everything holds so long as a kind of thermodynamic balance is preserved. When the aggregated resultant of components is altered, by virtue of internal tension or exchange, or by a combination of both, what is effected is what in thermodynamics is called 'a phase transition' – that is, a state change: what is determined is precisely a metamorphosis, that is, a relocating, a change of position or condition, a recombination that alters the whole through the formation of a new aggregate and a new equilibrium: form, territory,[48] potential dynamic. In sum, keeping in view that an entropic tension toward states of nonequilibrium (up to complete disorder) perpetually subsists, one might say that life exploits that tension to evolve, and withdraws from it, resisting in its process of formation and conservation, that is, by incorporating and incorporating itself, and preserving more or less linear ways. Life continually withdraws from states of equilibrium, yet never ceases to reproduce them – configuring itself as an unstable and variable mechanical order. In this regard, Erwin Schrödinger remarked: 'Life seems to be orderly and lawful behaviour of matter, not based

48 An additional aspect that would deserve to be addressed – though this has not been possible within the present scope – concerns the element of territory, or more precisely, the relation between *territory* and *soil*. A body incorporates the world first and foremost from its own act of localization, that is, by taking a place. Every organic state must be read, even before it is understood chronically, geographically: its taking-form is an acquiring of form within a specific territory; it is the construction of an *oikos*, which both determines and is determined by those who inhabit it. In this sense, organisms create a world – or rather, their own world – when they metabolize portions of the world by configuring territories. *Econthology*, therefore, cannot be conceived apart from the dynamics of territorialization.

exclusively on its tendency to go over from order to disorder, but based partly on existing order that is kept up', adding:

> *Living matter evades the decay to equilibrium.* [...] When is a piece of matter said to be alive? When it goes on 'doing something', moving, exchanging material with its environment, and so forth, and that for a much longer period than we would expect an inanimate piece of matter to 'keep going' under similar circumstances. When a system that is not alive is isolated or placed in a uniform environment, all motion usually comes to a standstill very soon [...]. After that the whole system fades away into a dead, inert lump of matter.[49]

Then it becomes clear that the category of *Being* is divested of centrality, remaining a fictive and operational device. For Nietzsche it is not a question of determining the fundamental ontological difference between *Being* and *beings*; rather, it is a question of defining the variability that is determined *echontologically* between nature and life, that is, between the two trends that characterize the world (exchanges, tendencies, impulses, alterations, interactions, and hence not 'beings'); we might conceive these two tendencies under the figure of a double Nature, or rather a Nature and its double, as already suggested by the passage from *The Gay Science*: a dual orientation toward equilibrium – that of life, and that of inanimate matter. These differ, once again, not by essence but only in a morphological and energetic sense, as energetic (thermodynamic) forms and states in ceaseless rearticulation. Such forms are both tending, although according to distinct 'grammars', toward chaos (or rather toward what Joyce called *chaosmos*) – that is, toward a form of irreversible equilibrium – in their slowings, in their accelerations, in their reconfigurations.

In fact, as is well known, whereas inorganic entropy is a constant process without internal resistance, organic entropy, as Schrödinger affirms, exhibits a different variability: he shows how 'organic matter' resists decay toward thermodynamic equilibrium by feeding

49 Erwin Schrödinger, *What is Life?* [1944] (Cambridge: Cambridge University Press, 1992), pp. 68–9.

on negative entropy. There is something in organisms that keeps them temporarily distant from a return to inorganic matter:

> It is by avoiding the rapid decay into the inert state of 'equilibrium' that an organism appears so enigmatic [...]. How does the living organism avoid decay? The obvious answer is: By eating, drinking, breathing and (in the case of plants) assimilating. The technical term is metabolism. [...] Every process, event, happening [...], everything that is going on in Nature means an increase of the entropy of the part of the world where it is going on. Thus a living organism continually increases its entropy – or, as you may say, produces positive entropy – and thus tends to approach the dangerous state of maximum entropy, which is death. It can only keep aloof from it, i.e. alive, by continually drawing from its environment negative entropy.[50]

We might therefore say, paraphrasing Nietzsche, that the living tend, through their irritable disposition, to perturb their own tendency toward entropic equilibrium – so that the progressive loosening of dynamic bonds, and the absence of a sufficient *quantum* of energy to reorganize them into a new aggregated form, does not lead the organism to its definitive decay. Here, then, we may begin to grasp in what sense, for Nietzsche, illness marks the transition from *potency* to mere preservation.

It is worth noting that twentieth-century biology and physics – though largely unaware of Nietzsche – have returned to this question in terms of *assimilation*, *nourishment*, and *dissipation*. It is particularly striking that, for instance, Schrödinger, within the trajectory from microscopic disorder to macroscopic order, posited the hypothesis that the 'phenomenon of life' might even follow laws of its own, distinct from but congruent with those of matter – and that such order would be linked to the capacity of organisms to maintain themselves 'by extracting "order" from the environment'.[51]

This too, then, is a matter of *Echontology*. In the end, we must ask whether Nietzsche, through the notion of *having*, did not suggest a new horizon for thought: one that no longer thinks in terms of equivalence, property, unity, and permanence, but in terms

50 Ibid., pp. 71–2.
51 Ibid., p. 73.

of appropriation, assimilation, exchange, incorporation. A new *morphology*, built upon the complex horizon of relational exchange, where *appropriation* and *incorporation* disrupt the dualism between *finalism* and *mechanism*: forms – organic and inorganic – move according to a combination of forces and interactions, and refine themselves by structuring as *incorporative impulses*. It is no longer a question of *what* a (living) body *is*, but rather *how far it can have*.

Antonio Carrano*

HOW FAR PHILOSOPHERS DARE
The Answer to Fears of the Future by F. C. S. Schiller

Abstract

This contribution examines Schiller's follow-up to the more famous texts of Haldane and Russell, who, in the mid-1920s, drew inspiration from the figures of Daedalus and Icarus to portray, from opposing perspectives, the unsettling post-war world. Appropriating the myth of Tantalus, the English leader of pragmatism offered a modernized version to give voice to a proposal for a reorganization of society in which biologism and evolutionism form the backdrop to a reflection that, aided by his adherence to Galton's eugenics, gives vent to a class-based philanthropy masquerading as humanism.

Keywords

Mythology – Science – Eugenics – Humanism – Pragmatism

1. In the mid-1920s – a time when Europe, fresh from the traumatic Great War, was already brewing even greater calamities – a group of prominent British intellectuals from varied fields sought to interpret the present through the lens of Greek mythology. John Burdon Sanderson Haldane, a biochemist and geneticist with Soviet sympathies, wrote a piece on Daedalus. In response, Bertrand Russell, a philosopher renowned for his liberal-democratic stance, penned one on Icarus, aiming to reignite the discussion about the

* Università degli Studi di Napoli Federico II.

ethical implications of scientific advancement. Finally, they were joined by the less familiar Ferdinand Canning Scott Schiller, an English champion of James's pragmatism, credited with a unique take on humanism that inspired his work, *Tantalus or The Future of Man*.

Without delving into the analysis of the first two texts, both products of conferences, we can begin with some pages from *The Scientific Outlook*, in which Russell alarmingly outlined the ethics of the future, shaping it around a reality molded by science. He acknowledged the effects produced by science 'as a technique'[1] on 'the forms of human society', and coupled this with a call for an 'increase in wisdom'[2] proportional to the 'increase in knowledge'.[3] His was the detached attitude of someone observing, before their very eyes, the evolution of a significant historical process: that of science's predominant 'attainment of power'[4] and the corresponding elimination of metaphysics 'as the attainment of truth'. Adhering rigorously to the scientific method, it was the scientists themselves, with their sceptical approach, who gradually undermined the value of 'science as metaphysics',[5] which the philosopher placed in that area bordering art and religion, where for a long time the 'madness of Prometheus' had taken root – that is, the ambition that inspired the greatest men's attempt 'to become gods'.

Just as the titan, in his mania, had embodied a 'value that is religious, not political, or even moral', as 'the only ultimate value of human life', so too did he who had forever chained him through calculating reason appear to him as the cold bearer of a new ethic, contrary to religion: an ethic which, being directed at disciplining collective rather than individual behavior, showed a direct connection with politics.

1 See Bertrand Russell, *The Scientific Outlook* (London: George Allen & Unwin LTD, 1919), p. 11.
2 Bertrand Russell, *Power: A New Social Analysis* (London: Allen and Unwin, 1938), p. 284.
3 See Bertrand Russell, *The Scientific Outlook*, p. 12.
4 For this and subsequent citations, see ibid., p. 104.
5 For this and subsequent citations, see ibid., p. 102.

> The new ethic which is gradually growing in connection with scientific technique will have its eye upon society rather upon the individual. It will have little use for the superstition of guilt and punishment, but will be prepared to make individuals suffer for the public good without inventing reasons purporting to show that they deserve to suffer. In this sense it will be ruthless, and according to traditional ideas immoral […].[6]

It's certain that, in identifying a trait of immorality, Russell intended to express his dissent towards the ethics of a world dominated by technology. Nevertheless, he couldn't move beyond a sterile critique of a phenomenon he conceived as the result of an inexorable process. For this reason, just as he would a few years later in his dispute with Haldane, he naively expressed himself in favor of the 'life of the emotions',[7] subordinating even knowledge to it, convinced that 'a world without delight and without affection is a world destitute of value'. What's more, he asserted the need to counter that ethics with a 'new moral outlook […] in which submission to the powers of nature is replaced by respect for what is best in man'. And this was not only because, realistically, 'where this respect is lacking scientific technique is dangerous', but because, ideally, science, besides emancipating man from the yoke of nature, should 'deliver him from bondage to the slavish part of himself'. To the 'fear' of the menacing dangers that loomed, he had nothing else to oppose but this: 'hope for the future', or rather, the wish that those dangers were not truly 'inevitable'. Truly too little.

It's not surprising that these final considerations conclude the last chapter of the book, which is dedicated to the relationship between science and values. Here, discussing the internal development of science, which had not yet reached completion, Russell characterized its trend as a 'passage from contemplation to manipulation'.[8] This transition saw the 'practical' impulse to dominate the world prevail over the 'contemplative' one, which aimed to understand it for its own sake. In his view, signs of this ascendancy could be

6 Ibid., p. 241.
7 For this and subsequent citations, see ibid., pp. 278–9.
8 For this and subsequent citations, see ibid., p. 270.

observed not only in the social and political spheres but also in the philosophical one.

> The power impulse is embodied in industrialism and in governmental technique. It is embodied also in the philosophies known as pragmatism and industrialism. Each of these philosophies holds, broadly speaking, that our beliefs about any object are true in so far as they enable us to manipulate it with advantage of ourselves. This is what may be called a governmental view of truth.

It is this accusation of complicity leveled by Russell against the manifold versions of pragmatism (of which James, Schiller, and Dewey, in his view, reflected the religious, literary, and scientific aspects, respectively)[9] that now offers us the cue for a brief consideration regarding the last of the protagonists in this short history: Ferdinand Schiller, James's English associate, now almost forgotten.[10] He intervened close on the heels of the dispute between Haldane and Russell, proposing another figure from Greek mythology as a model: Tantalus.

---

9 See Betrand Russell, *Sceptical Essays* (London: George Allen & Unwin, 1928), p. 60. In Chapter V, dedicated to twentieth-century philosophy, Russell began by discussing pragmatism, especially as formulated by James, caustically criticizing its theory of truth. He deemed it crude and cunning, likening it to the "practice of the propagandist (advertiser)" (or rather, the salesperson) (see ibid., p. 63). Even later, on the eve of the new war, Russell included pragmatism among the 'philosophies of power', again criticizing its conception of truth, which he saw as susceptible to being monopolized by dictators and their supporting social groups to consolidate their dominion (see Russel, *Power: A New Social Analysis*, pp. 272–3). Considering his rather superficial treatment of other philosophers (such as Fichte), who – despite holding strong positions, even if dictated by the political situation of their time – would have deserved a less superficial and forced exposition of their ideas, it must be said that pragmatism too is reduced to a simplified thesis that associates truth with material advantage, presenting it as pleasant.

10 A recent, albeit introductory, re-examination of his thought is offered by J. S. Pinilla, 'Una invitación a la lectura de Ferdinand Canning Scott Schiller', *ISEGORÍA. Revista de Filosofía Moral y Política*, 41(2009), 253–66.

2. In a text published that same year in the series where the other two writings[11] had already appeared, Schiller began by rejecting any accusation of pessimism, convinced that his reasoning had led to less bleak conclusions than those who preceded him. This was, if only because he did not consider 'the evils which threaten the future of humanity'[12] 'inevitable', to the extent that he outlined a plan to counter them. What others therefore branded as negative merely reflected a provident and responsible attitude, interested in identifying appropriate solutions; this stood in opposition to the unconscious stance of those who, if anything, continue to live as if nothing were wrong and head straight for the precipice, expecting 'to be saved by a miracle'. In rejecting the logic of opposites, he therefore identified Tantalus as a third figure, borrowed from mythology, to give voice to a position different from those who had sided with Daedalus as the champion of science and Icarus as its predestined victim.

Almost like in a fairy tale, Schiller resorted to the device of the dream, customary in writings intended to bear witness to extraordinary visions, imagining himself falling from the ruins of Tantalus's tomb (discovered at that time in Phrygia by English archaeologists), on which he had fallen asleep, only to find himself in the plain where, from time immemorial, the hero's torment was consumed. There, the tree of knowledge and the elixir of life are what the latter vainly reaches for, and which the philosopher, out of solidarity, tries in vain to help him attain until the moment of his own awakening.

The fleeting reference to Freudian psychoanalysis was merely a rhetorical artifice to exclude the help of others in finding the key to the dream, the interpretation of which thus takes shape in the continuation of the exposition, aimed at quickly tracing the lines of the course of history under the banner of the idea – intended to restore confidence – that the current situation reveals nothing new,

11 John Burdon Sanderson Haldane, *Daedalus, or Science & the Future* (New York: E. P. Dutton, 1924), Betrand Russell, *Icarus or The Future of Science* (New York: E. P. Dutton, 1924).

12 For this and subsequent citation, see Ferdinand C. S. Schiller, *Tantalus or The Future of Man* (New York: E. P. Dutton, 1924), pp. v–vi.

since the 'human race' has always lived in the precarious condition of constantly threatening its own destruction and at the same time avoiding it.

Against this backdrop, Schiller undertook a personal reconstruction of 'the history of Man',[13] first noting the 'arrest of his biological development' – 'certainly the greatest mystery' – which he attributed to the emergence of a society organized around collaboration. This collaboration negatively altered, and thus distorted, 'the control of natural selection'.[14]

> Now, if we suppose that by midpalæolithic times man had established his ascendency over nature and perfected his social organization sufficiently to render these services to his fellows, we have suggested a possible cause of the cessation of biological progress. For social influences are as likely as not to be 'contra-selective', that is, to tend to preserve by preference the stocks which are less viable from a merely biological point of view. They are markedly so at present, and it would be asking too much to expect the tribal chiefs of early men to have been wise and provident enough to see to it that their social institutions were eugenical in their effects.[15]

For Schiller, it was undeniable that the arrest of biological selection corresponded to the progress of Man as a social and cultural being. This progress was critically dependent on the inventions of *Language* and *Writing*, which he saw as the basis of the 'vast

13 For this and subsequent citation, see ibid., pp. 15–16.

14 It can't truly be said that Schiller's thesis was original or new, given that traces of it can even be found in a 1864 article by Wallace, appreciated by Darwin himself, where it states in a not dissimilar manner (Alfred Russel Wallace, 'The Origin of Human Races and the Antiquity of Man deduced from the theory of "Natural Selection"', *The Anthropological Review, or Journal of the Anthropological Society of London* (London: Trübner, 1864), vol. II, pp. clxiii–clxix). It must be said, however, that Wallace's analysis was characterized by an optimistic vision, converging with the Spencerian concept of evolution as a process tending to achieve an advanced and thus also evolved society, whose image diverged greatly from that defined by Schiller, who was convinced of the ongoing decadence of European civilization.

15 Schiller, *Tantalus or The Future of Man*, pp. 17–18.

intricate structures, religious, political, social, and scientific, which knit together and dominate human societies from generation to generation'.[16] Similarly, it was undeniable that human institutions were physiologically destined 'to get clogged with the waste products of their own working', eventually 'becoming the most formidable instruments for their frustration'. This undermined confidence in progress itself, eroding the reasons for pride in civilized Man: to all intents and purposes, 'the irrational, impulsive, emotional, foolish, destructive, cruel, credulous, creature he always was'.[17]

'The painful truth is that civilization has not improved Man's moral nature':[18] this was Schiller's conclusion to be drawn from an honest look at 'the history of the past fifty years'. This history, on one hand, presented an image of Man as a civilized individual, yet ready to unleash his crude instincts if something upsets the established social order. On the other hand, it exposed everyone to a growing sense of precariousness, a consequence of the rapid acceleration of the transformation process. Regarding this, the philosopher could not help but warn: 'it is risky to expose the inelastic nature of so stubbornly conservative a creature to new conditions at a rapid rate';[19] it is risky to expose 'the palaeolithic savage masquerading in modern garb to a series of physical and mental shocks' that ultimately 'have endangered his equilibrium'. He further noted: 'Under these conditions, the fighting instinct ceases to be an antiquated foible, like the hunting instinct, and becomes a deadly danger'.

Speaking in this manner, Schiller still exhibited olympic detachment. However, it would not take much for the cold observation to give way to a more invested analysis. This indeed happened when he indulged in a partisan social reflection, marred by his adherence to questionable eugenic positions, which were widely diffused and scarcely divisive

16 For this and subsequent citation, see ibid., p. 20.
17 See ibid., pp. 31–2.
18 For this and subsequent citation, see ibid., pp. 32–3.
19 For this and subsequent citation, see ibid., pp. 34–5.

politically at the time.[20] In fearing a 'degeneration of mankind',[21] he too fell into a 'classist' line of reasoning, consistent with that of Ronald A. Fisher, the future author of *The Genetical Theory of Natural Selection* (1930). Fisher, a colleague of Haldane at University College London (UCL) and, like other eugenicists, fascinated by the history of Rome and other classical civilizations, attributed their decline to the prolificacy of the common people. Sharing the idea that an 'inverse relationship exists between fertility and "social value"',[22] Schiller expressed himself thus in *Tantalus*:

> A simple and easily observable sociological fact is at the bottom of mischief. The different classes in a society have different birth-rates and death-rates, and the differences between these yield their several net rates of increase or decrease. Now, whereas under conditions of savage life class differences can hardly exist, or, at least cannot be accentuated, so that the whole tribe flourishes or perishes together, and among barbarians the upper classes have a very great advantage and the tribe recruits itself chiefly from the children of the chiefs, because the conditions of life ae so severe that the lower classes are not able to rear many children; in civilized societies these conditions are reversed. It is found that though both birth-rates and death-rates grow as we descend the social scale, so does the net rate of increase.[23]

Even without translating into Fisher's political program, which hypothesized fiscal interventions to support the middle classes and social policies aimed at disadvantaging larger families, Schiller's reflection was thus rooted in that area of thought which saw the demographic decrease of 'the highest classes'[24] as the danger that 'chief civilized societies' were inexorably facing. Inclined by their tendency to favor a policy of social promotion with predictably disastrous effects, due to the fact that it would ultimately produce an

20 Cf. Geoffrey Russell Searle, *Eugenics and Politics in Britain 1900–1914* (Leyden: Noordhoff International Publishing, 1976), chap. 2, 'The Development of a Eugenics Movement in Britain', pp. 9–19.

21 See Ferdinand C. S. Schiller, *Tantalus or The Future of Man*, p. 36.

22 Cf. Adam Rutherford, *Control: The Dark History and Troubling Present of Eugenetics* (London: Weidenfeld & Nicolson, 2022).

23 Ferdinand C. S. Schiller, *Tantalus or The Future of Man*, pp. 37–8.

24 For this and subsequent citation, see ibid., p. 39.

increase in demographic disparity between the different classes, all to the benefit of the popular strata, coupled with a growing softness of the most advanced nations, such societies were condemned, in his opinion, to a sure degradation, the final outcome of which was not difficult to predict.

> It is superfluous to insist either on the fatuity of a social organization such as this, or on the certainly of racial degeneration which it entails: but it may be well to draw attention to the rapidity with which these degenerative processes are at the present sapping the vitality and value of our civilized races. The failure to reproduce does not, as in former times, affect merely the aristocracy in the highest social strata; it has spread to the whole of the professional and middle classes, and to most classes of skilled labour.[25]

Schiller did not, in fact, refrain from enumerating the causes of what appeared to him to be an authentic 'class-discrimination':[26] a decrease in mortality consequent to advances in medicine, improvement in hygienic conditions, increased benevolence towards the lower classes, progress in obstetrics, and finally, social promiscuity. Convinced that he should not succumb to pessimism,[27] he did not, however, limit himself to listing the reasons for gradual decadence, instead indicating the two possible strategies aimed at curbing it and reversing its trend. Between a 'moral reform',[28] capable of drawing on authentic 'Christian ethics' (if only for reasons of survival, considering the greater adaptability of 'the human "rabbit"' compared to 'the Nietzschean "wild beast"'), and a 'eugenical reform',[29] accompanied by the 'reconstruction of our social organization', he had no doubts about which was the more

---

25 Ibid., pp. 41–2. Cf. Dan Stone, *Breeding Superman: Nietzsche, Race and Eugenics in Edwardian and Interwar Britain* (Liverpool: Liverpool University Press, 2002), chap. 4, 'Race and Eugenics', pp. 94–114.

26 See ibid., p. 43.

27 Schiller had extensively discussed pessimism in his *Humanism: Philosophical Essays* (London: MacMillan, 1903), 'The Place of Pessimism in Philosophy', pp. 157–65.

28 For this and the subsequent quotations, see ibid., pp. 48–9.

29 For this and the subsequent quotations, see ibid., pp. 51–2.

effective. While rejecting the idea of introducing eugenics 'from above', in an imperious form, 'by way of administrative order', and emphasizing the experimental nature of its practice, open to the correction of errors, he imagined that one could try, by common agreement, 'to improve and increase the best types':[30] without proceeding 'upon any cut-and-dried scheme', and indeed allowing oneself to be guided 'by the results of experiment and the fruits of experience', made the subject of public discussion.

To compensate for the 'contra-selection'[31] operated by civilization until it was annulled, to intervene in a balanced way 'by the elimination of defectives at the bottom of the social scale, and by the increase of ability at the top', with a view to recovering the 'plasticity' and 'progressiveness' once manifested by man[32] – this is what Schiller believed eugenics could do well. At least until Psychology became 'an effective science',[33] so as to obtain 'some surprising effects [...] from the actual human mind'. While waiting, therefore, to be able to act directly on the latter, working on behavior, he thought that one could in the meantime intervene on bodies, seeing in eugenics a ready-to-use remedy against the pressing degeneration.

It might be said that if the idea that the fate of humans isn't sealed forever can ward off pessimism, the remedy Schiller devised to revitalize the society of his time couldn't lift it from a bleak vision of the future. Nor could it reassure those who, in his opinion, should have experienced it: men like the philosopher, to whom the moral he drew from the story of Tantalus – as it was presented to them – offered nothing good, having everything to fear from the proposal to restore the healthy mechanism of natural selection in the interest of the socially 'highest classes'. Moreover, although abstractly meritorious, the decrease in the defects of some accompanied by the increase in the merits of others was not conceived to produce a genuine leveling

30 For this and the subsequent quotations, see ibid., pp. 58–9.

31 For this and the subsequent quotations, see ibid, p. 60.

32 Referring to *Tantalus*, Schiller made similar observations in a later work, *Logic for Use. To the Voluntarist Theory of Knowledge* (New York: Harcourt, Brace & Co., 1930), chap. 10, 'The Biologic Judgement', p. 196.

33 For this and the subsequent quotation, see ibid., p. 63.

of everyone's conditions in practice, but rather to sterilize the social environment in which the class divide had to remain unchanged, in the absence of a mechanism aimed at producing a fairer distribution of resources. For this reason, if the character of Tantalus embodies the man desirous of goods that constantly elude him, forgetting the popular saying that advises moderation ('grasp all, lose all'), it's unclear why, by actualizing it, Schiller didn't aim to highlight the harm that comes from unrestrained ambition and greed. By depriving the text of any reference to the dynamism and instability of the capitalist social system, found in one way or another in the writings of Haldane and Russell, he became a proponent of a purely technical program of social reorganization, to which 'humanistic' pragmatism came to his aid with its utilitarian manner of legitimizing values.

3. One can now see in the reform proposed by Ferdinand Schiller a response to the phenomenon of decadence that Francis Galton had warned about since the second half of the nineteenth century. A typical example of that 'intellectual aristocracy' mentioned by those who, in a recent reconstruction of the complex history of eugenics,[34] have sought to emphasize its class roots, the author of *Hereditary Genius* was the first – as is well known – to be convinced 'that each generation has enormous power over the natural gifts of those that follow',[35] recognizing also the 'duty we owe to humanity to investigate the range of that power, and to exercise it in a way that, without being unwise towards ourselves, shall be most advantageous for the future inhabitants of the earth'.

---

34 Cf. Carlo Alberto Defanti, *L'eugenetica: un tabù contemporaneo. Storia di un'idea controversa* (Turin: Codice Edizioni, 2012), p. 19.

35 For this and the subsequent quotations, see Francis Galton, *Hereditary Genius* (London: Macmillan, 1869), p. 1. As he would express it a few years later, eugenics was to be conceived as a practice aimed at pursuing 'the ends of evolution more rapidly and with less distress than if events were left to their own course' (see Francis Galton, *Inquiries into Human Faculty and its Development* (London: Macmillan, 1883), p. 2), and thus favoring rather than contrasting 'the course of evolution' (see ibid., p. 334), whose doctrine showed a genuine 'religious significance', such as to suggest 'an alteration in our mental attitude' and to impose 'a new moral duty' (see ibid., p. 337).

However, what interests us here is not the 'classist' imprint of the superficial philanthropy expressed by Schiller at the end of the 1920s. Without forgetting, for this reason, the anti-democratic trait of his political orientation that underlies his positions,[36] and which led him to be culpably lenient with Fascism,[37] it is rather important to ultimately clarify if and how his 'despotic' approach as a pragmatist (to repeat Russell's trenchant judgment) to the problem of defining truth could have aligned with the humanism he promoted in the philosophical field, in conjunction with his adherence to pragmatism.

It is worth recalling: it was in a text of the same name from the early twentieth century that Schiller – after primarily crediting pragmatism with overcoming 'the old antithesis of Faith and Reason',[38] in terms that closely resemble James's[39] – identified humanism as the 'final

36 Schiller, in fact, dedicated a good number of essays to eugenics, subsequently collected in two volumes: *Eugenics and Politics* (London: Constable, 1926) and *Social Decay and Eugenical Reform* (London: Constable, 1932). Convinced that it constituted an 'application of biology to social life, as a sort of social hygiene on a large scale', and at the same time that 'progress is not a necessity either of thought or of nature', considering the 'always [...] imminent danger of degeneration' (see *Eugenics and Politics*, pp. 1–3), he believed 'a program of "positive" eugenics' to be 'practicable [...] here and now, with quite moderate and simple means', capable of influencing '(if persisted in and not suppressed by violence) [...] progressively the human race, in such a way as to improve its intrinsic qualities and render it more and more capable of improving its conditions of life' – even to 'reverse the processes of social decay which are now becoming so painful and so unmistakable'. See *Social Decay and Eugenical Reform*, p. v.

37 See in this regard Reuben Abel, *The Pragmatic Humanism of F. C. S. Schiller* (New York: King's Crown Press /Columbia University, 1955), pp. 145–7.

38 For this and the subsequent quotation, see *Humanism: Philosophical Essays*, p. xiv. The philosopher revisited the topic years later in *Must Philosophers Disagree? And Other Popular Essays in Popular Philosophy* (London: MacMillan, 1934), Chapter XXIII, 'Pragmatism, Humanism and Religion', pp. 306–19.

39 James, in turn, recognized the philosophical value of Schiller's positions, dedicating a chapter of his most extensive text on pragmatism to him, titled 'Pragmatism and Humanism'. See *Pragmatism: A New Name for*

theory of life' to be superordinate to pragmatism itself, because it responded to the interest of every human being.[40] It's not entirely clear, however, what he more specifically meant by this term. In fact, while on one hand the philosopher grounded it theoretically in a form of empiricism, acknowledging its merit in moving from 'human experience'[41] (as a 'clue to the world of human experience'), instead of relying on the 'intellectual game' of determining 'infinite possibilities of conceptual rearrangement', discernible in 'attempts to construct experience *a priori*', on the other hand, he emphasized its practical projection in the name of the 'view that the universe is ultimately "a joint-stock affair"' – thus reinterpreting Protagoras's ancient saying 'that Man is the measure of all things', starting with the 'whole experience-world', in a social key.[42]

As is well known, Schiller established a close integration between Pragmatism and Humanism, despite their subordinate relationship. Indeed, given the role of a mere 'method'[43] assigned to pragmatism, functional to the determination of the nature of truth[44] – for which

---

some Old Ways of Thinking. Popular Lectures on Philosophy (London, Bombay and Calcutta: Longmans, Green, 1907), pp. 239–70. Within it, while acknowledging the *in fieri* nature of truth (like language and law), James could not hide some perplexity regarding how, after reducing them to human products, Schiller then asserted the 'plastic' nature of the entire world (see ibid., p. 244).

40 On this topic, cf. Mark J. Porrovecchio, *F. C. S. Schiller and the Dawn of Pragmatism. The Rhetoric of a Philosophical Rebel* (Lanham–Boulder–New York–Toronto–Plymouth, UK: Lexington Books, 2011), Chapter 5, 'All too Human: The Flagging Relevance of Humanism', pp. 202–11.

41 For this and the subsequent quotations, see Schiller, *Humanism: Philosophical Essays*, pp. xix–xx.

42 Cf., in this regard, Porrovecchio, *F. C. S. Schiller and the Dawn of Pragmatism*, pp. 99–140.

43 For Schiller, what specified Pragmatism is 'the thorough recognition that the purposive character of mental life generally must influence and pervade also our most remotely cognitive activities. In other words, it is a conscious application to the theory of life of the psychological facts of cognition as they appear to a teleological voluntarism'. Schiller, *Humanism: Philosophical Essays*, p. 8).

44 'Pragmatism, on the other hand, essays to trace out the actual 'making of truth', the actual ways in which discriminations between the true and the

utility serves as a principle of selection and a criterion for validation, in line with the need for its social recognition[45] – it was then up to humanism to identify the authentic interest in relation to which the 'real truth and validity' of every assertion could be verified, as it aimed at defining the ultimate vision of life. For this reason, it was not only the pragmatist but also the humanist who spoke, stating that 'human interest [...] is vital to the existence of truth'.[46] For both, the equivalence between the true and the good was taken for granted – a truism, so to speak – assuming that both were to be evaluated, in his opinion, in relation to their consequences; furthermore, in relation to '"consequences" *to* some one and *for* some purpose'.

> If it is clearly grasped that the 'truth' with which are concerned is truth *for man* and that the 'consequences' are human too, it is, however, really superfluous to add either (1) that the consequences must be *practical*, or (2) that they must be *good*, in order to distinguish this view sharply from that of rationalism. For (1) all consequences are 'practical', sooner or later, in the sense of affecting our action. Even where they do not immediately alter the course of events, they alter our own nature, and cause its actions to be different, and thus lead to different operations on the world. Similarly (2) if an assertion is to be valuable, and therefore 'true', its consequences must be 'good'. They can only test the truth it claims by forwarding or baffing the interest, by satisfying or thwarting the purpose, which led to the making of

false are effected [...]'. Ferdinand C. S. Schiller, *Studies in Humanism* (London: MacMillan, 1907), p. 4). As stated later, '[p]ragmatism will seem a special application of Humanism to the theory of knowledge. But Humanism will seem more universal. It will seem to be possessed of a method applicable universally, to ethics, to aesthetics, to metaphysics, to theology, to every concern of man, as well as to the theory of knowledge'. Ibid., p. 16.

45 Schiller had already expressed himself years earlier, in his theoretically strongest work, emphasizing the role of the influence of human social nature in the affirmation of a truth, whose premise remains, however, 'the responsibility of enunciating it' on the part of someone in one of the many abstract ways available to enunciate it. Ferdinand C. S. Schiller, *Riddles of the Sphinx. A Study in the Philosophy of Humanism*, 2nd ed. (London: Swan Sonnenschein, 1910), pp. 132–3.

46 For this and the subsequent quotations, see Schiller, *Studies in Humanism*, pp. 5–6.

> the assertion. If they do the one, the assertion is 'good', and *pro tanto* 'true'; if they do the other, it is 'bad' and 'false'. For whatever arouses an interest or forwards an end is judged to be (so far) 'good', whatever baffles or thwarts is judged to be 'bad'.

It's not the place here to ask in what sense the indication of 'the evaluation of claims' can serve as a 'Humanist test',[47] as proposed by Schiller. This would inevitably imply a more accurate discussion of the various definitions he provided for pragmatism. Regarding this, we can merely observe that the evolutionist premise supporting it – namely, that '[t]he whole mind, of which the intellect forms part, may be conceived as a typically human instrument for effecting adaptation'[48] – reconciles with the demand for concreteness required by humanism. This demand, when considering the 'real problems of life',[49] insists on '*leaving in* the whole rich luxuriance of individual minds, instead of compressing them all into a single type of "mind", feigned to be one and immutable', and therefore on addressing them while preserving 'the psychological wealth of every human mind and the complexities of its interests, emotions, volitions, aspirations'.

The impression is that this claim of concreteness in his philosophy, while respecting human complexity, prevented him from adequately justifying its theoretical foundations. These foundations were, in terms of objectivity, compromised by the appeal to the multitude of 'interests, emotions, volitions, aspirations', from which even the philosopher does not escape. But perhaps, in the conviction of asserting simple truisms, Schiller did not reflect enough on the consequences to be drawn from the application of his much-vaunted 'Humanist test'.

If consistency dictates that whoever sets binding conditions should be the first to adhere to them, it doesn't seem Schiller followed this principle. In fact, after ruling out the possibility of seeking truth in the abstract (as intellectualism, in his view, proposed), and convinced that utility determines it within the limits imposed by

47 See ibid., p. 1.
48 See ibid., p. 10.
49 For this and the subsequent quotation, see ibid., p. 13.

the interest one holds, he disregarded this premise by presuming the *a priori* validity of his own theses. Even if he had wanted to, he could not have strongly legitimized them on a theoretical level, once it was assumed that, after all, it is always someone's interest (including his own) that prevails, furthermore asserted for some purpose (for example, the moral reform hypothesized to prevent 'the evils which threaten the future of mankind').[50] The commitment to keep reflection 'within the limits of actual knowledge'[51] – in contrast to the tradition of metaphysics, bound by a criterion of 'coercive and objective validity' – obliged him to 'recognize all realities, [...] employ all conceptions and methods that [are] deemed necessary or expedient, without affirming their ultimate validity'.

That Mill was indicated by James as one of the guiding lights of pragmatism is evidenced by the dedication James affixed to his text of the same name, published in the same year as Schiller's *Studies in Humanism*, 1907. However, that the utilitarianism espoused by Mill in the moral sphere, concerning individual and collective behavior, was reinterpreted in a theoretical sense by Schiller is demonstrated by the equation he established between truth and utility, which was, in turn, the result of the hasty association of a thesis's proof with its application.

> To become really true it [a truth] has to be tested, and it is tested by being *applied*. Only when this is done, only that is when it is *used*, can it be determined what it really means, and what conditions it must fulfil

50 Schiller, *Tantalus or The Future of Man*, p. v. On the contextual and subjective nature of truth, cf. Abel, *The Pragmatic Humanism of F. C. S. Schiller*, Chapter 8, 'The Making of Truth', pp. 93–109, and Kenneth Winetrout, *F. C. S. Schiller and the dimension of pragmatism* (Columbus: Ohio State University Press, 1967), Chapter 5, 'Schiller's philosophy as subjectivism', pp. 107–37, have previously drawn attention.

51 For this and the subsequent quotation, see Schiller, *Studies in Humanism*, p. 17. It should be noted, however, that Schiller's own consideration of the history of philosophy is conditioned by a personalistic vision that led him to see it as 'an eloquent paean to the triumph of the personality' of various thinkers. Ferdinand C. S. Schiller, *On Human Truths* (New York: Columbia University Press, 1939), pp. 178–9.

> to be really true. Hence all real truths must have shown themselves to be useful; they must have been applied to some problem of actual knowing, by usefulness in which they were tested and verified.[52]

If not all truths are equal, not requiring validation in the same way, then Schiller's error was precisely in standardizing the concept of truth by defining it in relation to use. If this were the case, then none of the statements made in *Tantalus* or the other two texts dedicated to eugenics, and proposed as valid by their author, could be assumed *a priori* as such before being put to the test. Not only that, but if we admit that the 'value of consequences' is determined for some purpose and by someone's interest, then we must deduce that those statements are not imposed based on principles, but are confirmed only in the eyes of those who have good reasons to believe in them, striving to achieve objectives that the majority of people would not be willing to accept and adopt.

This ultimately shows how the term 'application' was used in a dual sense: not only in relation to the justification of truths destined to appear 'real' because also 'useful', thereby modifying the 'conception of reality',[53] but also concerning prospective reasoning aimed at formulating a program, with organizational implications, for the regeneration of society. A program that, indeed, excluded the use of the most advanced knowledge in biology and pharmacology[54] in the name of the 'pragmatic test of successful working',[55] preferring to look to 'pragmatically efficient Psychology', destined one day to 'invert the miracle of Circe, and really transform the Yahoo into a man'. But which, in the meantime, proposed, with a view to

52 Ibid., p. 8.

53 See ibid., Chapter 7, 'The Making of Truth', p. 192.

54 'For this purpose, I am much less inclined to put my trust in the advances of pharmacology than Mr. Haldane and Russell (cf. *Daedalus*, p. 34; *Icarus*, p. 54). Hitherto new drugs have only meant new vices, sometimes (like cocaine) of so fascinating a character as to distract the whole police force from their proper function of repressing crime. So it seems legitimate to be very sceptical about moral transformation scenes to be wrought by pills and injections' (Schiller, *Tantalus or The Future of Man*, pp. 61–2).

55 For this and the subsequent quotations, see ibid., pp. 63–4.

collective prophylaxis guided by prudence and moderation, 'a safer though slower way, that of the eugenical reform and reconstruction of our social organization'.[56]

56 Ibid., p. 51. Recalling the positions expressed by Haldane and Russell, Schiller continued: 'As to the possibilities in this direction, I incline to be much more hopeful than Mr Haldane or Mr Russell. Mr Haldane despises eugenics, because he is looking for the more spectacular advent of the "ectogenetic baby", to be the Saviour of mankind. But he may not arrive, or be seriously delayed in transmission, or fail to come up to Mr Haldane's expectations; and, meanwhile, we cannot afford to wait. Mr Russell distrusts eugenics, because he fears that any eugenical scheme put into practice will be "nobbled" by our present ruling rings, and perverted into an instrument to consolidate their power. He thinks that dissent from dominant beliefs and institutions will be taken as proof of imbecility, and sterilized accordingly (*Icarus*, p. 49), and that the result would merely be to spread over all the world the hopeless uniformity and commonplaceness of the ideals and practice of the American business man, as depicted by Mr Sinclair Lewis. This prognostication would be very plausible, if we supposed eugenics to be introduced into the social structure from above, privily, and in small doses, and by ways of administrative order, as under the existing Act to check the spread of feeble-mindedness. But this method would be impracticable. It would not generate anything like the social momentum necessary to carry through any radical reform. To make it effective, it would have to be backed by a powerful, enthusiastic, and intelligent public sentiment. This presupposes that the public has been biologically educated to appreciate the actual situation, and has been thoroughly wrought up about the fatuity of our social order, and understands what is wrong with it' (ibid., pp. 51–3).

THE AUTHORS

Giulia Abbadessa is a research fellow in French literature at the University of Florence and a PhD candidate in Philosophical Sciences at the University of Federico II. She previously earned a European PhD in Languages, Cultures, and Comparative Literatures from the University of Florence, the Sorbonne, and the University of Bonn. She has also collaborated with the Leopardi Laboratory at Sapienza University and with the Italian National Research Council (CNR). Her research focuses on mythology and the interaction between literature and philosophy.

Blaise Bachofen is an Associate Professor (HDR) in Philosophy at the Faculty of Law of CY Cergy Paris Université and at Sciences Po Saint-Germain-en-Laye. He specializes in Enlightenment philosophy, modern political thought, and legal theory. His most recent publications include *La philosophie de Rousseau* (Vrin, 2024), and the co-editing of issues 44 and 45 of the journal *Lumières*, devoted to the theme *The 'Savages' of the Enlightenment* (2024/2 and 2025/1). Forthcoming: *Rousseau, éléments d'une théorie de la guerre* (PUF, 2026).

Andrea Bocchetti is a researcher in the History of Philosophy at the Federico II University of Naples. He has previously taught courses in philosophy at the Institut Catholique de Paris and at Sorbonne Université. His research focuses on modern philosophy, epistemology, and the metaphysics of subjectivity, with particular attention to Descartes, Vico, and Nietzsche. He is a member of the Italian Society for the History of Philosophy, and a collaborator of the Centre d'études cartésiennes at Sorbonne Université. He has authored two monographs and numerous articles on Descartes, Nietzsche, and Vico, published in journals such as *Giornale di metafisica*, *Les études philosophiques*, and the *Bollettino del Centro di studi vichiani*. He also serves on editorial boards and coordinates research collaborations between the Istituto Italiano per gli Studi Filosofici and leading French Universities.

Raffaele Carbone is associate professor at the Department of Humanities of Federico II University of Naples, where he currently directs the PRIN research project 'Paradoxical Humanism and Environmentalism' and the Laboratory 'Humanism, Arts and Social Research'. He was also 'Directeur de programme' and vice president of the Assemblée Collégiale at the Collège International de Philosophie. He has carried out extensive research at various institutes of advanced study and research centres in France (CHSPM/Université Paris 1 Panthéon-Sorbonne, Collegium de Lyon, FMSH, IEA de Nantes) and Germany (Centre Marc Bloch) and has taught at the universities of Tours and Montpellier 3. His research focuses on early modern philosophy and its influence on contemporary authors, as well as on critical theory. His publications include *'Différence' e 'mélange' in Montaigne* (Mimesis, 2013), *La Vision politique de Malebranche* (Classiques Garnier, 2018), the special issues on Spinoza of *Rivista di Storia della filosofia* (3/2023), the edited volume (in collaboration with Katia Genel) *Avant le crepuscule: Réinterroger Horkheimer et l'histoire de l'École de Francfort* (Classiques Garnier, 2026).

Antonio Carrano is full professor of History of Philosophy at the Department of Humanities of Federico II University of Naples. Author of essays and monographs on topics and authors of the German philosophy of the eighteenth and nineteenth centuries, he has translated and edited works by Wilhelm von Humboldt and Fichte.

Armando Mascolo is Senior Researcher in the History of Philosophy at the Institute for the History of Philosophy and Science in Modern Age (ISPF) of the National Research Council (CNR) in

Naples. His interests are mainly focused on the study of European philosophical thought between the nineteenth and twentieth centuries, with particular attention to irrationalist and vitalist philosophies, existential themes, and the issues of nihilism and the tragic. He directs *Rocinante. Rivista di filosofia iberica, iberoamericana e interculturale* and has published studies on Leopardi, Ganivet, Unamuno, Ortega y Gasset, Sartre, Cioran, Rensi and Zubiri.

Ph.D. in Philosophy (Université Paris 1 Panthéon-Sorbonne / Universidade de São Paulo), LEONARDO O. MOREIRA is Associate Professor of Modern and Contemporary Political Philosophy in the Department of Philosophy at Université Paris 8. He is also a researcher at the Laboratoire d'études et de recherches sur les logiques contemporaines de la philosophie (LLCP) and an associate member of the Centre d'histoire des philosophies modernes de la Sorbonne (HIPHIMO). His publications include *Jean-Jacques Rousseau, Variations sur le sauvage* (Classiques Garnier, 2025), the co-edited collective volume *Os selvagens de Rousseau* (São Paulo, Phi, 2021) as well as the thematic double issue 'Les "sauvages" des Lumières' in the journal *Lumières* (44–45, 2024/2025).

STEFANO PETRUCCIANI is Honorary Professor in Political Philosophy at the University of Rome La Sapienza. He was President of the Italian Association for Political Philosophy from 2013 to 2019. He is author of many books including: *Modelli di filosofia politica* (Einaudi 2003), *Democrazia* (Einaudi 2014), *Politica. Una introduzione filosofica* (Einaudi 2020), *The Ideas of Kark Marx* (Palgrave 2020), *Theodor W. Adorno's Philosophy, Society and Aesthetics* (Palgrave 2021), *Pensare con Marx* (Carocci 2022), *Marx en 10 palabras* (Alianza Editorial 2023), *La Scuola di Francoforte* (Carocci 2023). He edited a history of marxism in three volumes (Carocci 2015).

ANNA PIA RUOPPO is associate professor of Moral Philosophy at the University of Naples Federico II and collaborates as an Erasmus lecturer with the Technische Universität in Darmstadt. She is also a member of the doctoral college in Philosophy at the Universidad Alberto Hurtado de Santiago de Chile, an external collaborator at the

Praxis Research Centre in Portugal, and a member of the Centre for Studies on Marx and Marxism based at the University of Padua, and of the HUMANISE ('Humanism, Arts and Social Research') based at the Department of Humanities at the University of Naples Federico II. She studied in Germany (Heidelberg, Tübingen, Freiburg i.Br.) and France (Paris). She was a post-doctoral fellow at the University of Salerno and obtained a PhD in Philosophical Sciences (Naples 2004) and one in Ethics and Political-Legal Philosophy (Salerno 2011). She is the author of numerous essays and translations, as well as the books: *Vita e Metodo* (Florence, 2008), *L'attimo della decisione* (Genoa, 2011), *Pensare l'esistenza all'altezza della prassi* (Naples, 2022) and *Marxismo ed esistenzialismo: due filosofie dell'Europa – Lukács e Jaspers si incontrano a Ginevra (1946)* (Milan, 2023).

François Specq is Professor of American Literature and Culture at the Ecole Normale Supérieure de Lyon, France, and a researcher affiliated with IHRIM (in the Centre National de la Recherche Scientifique). His scholarly activities focus on American literature from the colonial era to the early twentieth century and on ecocriticism and environmental humanities. Among his publications are critical studies and translations of works by Thomas Jefferson, Henry David Thoreau, Ralph Waldo Emerson, Herman Melville, Frederick Douglass, Margaret Fuller, Mary Austin, and Jack London, and edited collections including *Thoreauvian Modernities* (University of Georgia Press, 2013), *Walking and the Aesthetics of Modernity* (Palgrave Macmillan, 2016), *Thoreau Beyond Borders* (University of Massachusetts Press, 2020), and *Pensées de l'architecture aux États-Unis/Thinking Architecture in the United States* (Paris: Belin, 2025).

Friedemann Stengel, born in Eisenach/former GDR, studied Protestant Theology in Halle from 1986 and, after the fall of the Berlin wall, in Bonn until 1992, was doctoral student in Contemporary History in Halle and in Berlin and received his doctorate in 1997 with a thesis on the theological faculties as a problem of the church and university policy of the communist party in the GDR during the Cold War. Around 2000 he was Research Assistant in Halle, Jena, and Leipzig for some years. From 2004 to 2010 he was member in the research group '*The Enlightenment* in the Referential Context of *Modern Esotericism*' funded by the German Research Foundation and located at the Interdisciplinary Centre for European Enlightenment Studies at Halle University. In 2009 he habilitated at the University of Heidelberg and was appointed to the Chair of Modern Church History at the Faculty of Theology in Halle in 2010, initially as a deputy professor and a few years later as a full professor. From 2016 to 2021 he was Scientist in Charge of the ITN project 'The History of Freedom and Human Dignity in Western Civilization', funded by the European Commission, and supervised English-language dissertations by two Italian doctoral students. He has been Director of the Interdisciplinary Centre for Pietism Research since 2018.

Irene Viparelli holds a PhD in Ethics and Political-Legal Philosophy. She is a researcher at the University of Évora in Portugal and a member of the Centro de Investigação em Ciência Política (CICP). She is the author of the book *'Oltre i limiti di Marx'. Un confronto tra Negri e Althusser* (Mimesis, 2017), she has also published numerous articles on topics related to Marxist theory, Italian workerism, and Althusserianism.

Roberta Visone is a researcher and lecturer of History of Philosophy at the Department of Humanistic Studies at the University of Naples Federico II. Her research interests are centred on the history of British and Italian philosophical and scientific thought between the second half of the nineteenth century and the beginning of the twentieth century. She is the author of the book *Selezione naturale ed equilibrio mobile della natura. L'evoluzionismo di Alfred Russel Wallace tra Darwin e Spencer* (Naples, 2022) and the monograph on Herbert Spencer's thought *Prima dell'evoluzione. Le radici politiche della filosofia di Spencer e la Social Statics del 1850* (Florence, 2010). She has published several essays on Carlo Michelstaedter, Herbert Spencer, Alfred Russel Wallace and the themes of 19th century evolutionism. She has also edited the Italian translation of Herbert Spencer's *The proper sphere of government* (Rome, 2013).

INDEX

MIMESIS GROUP
www.mimesis-group.com

MIMESIS INTERNATIONAL
www.mimesisinternational.com
info@mimesisinternational.com

MIMESIS EDIZIONI
www.mimesisedizioni.it
mimesis@mimesisedizioni.it

ÉDITIONS MIMÉSIS
www.editionsmimesis.fr
info@editionsmimesis.fr

MIMESIS COMMUNICATION
www.mim-c.net

MIMESIS EU
www.mim-eu.com

www.ingramcontent.com/pod-product-compliance
Lightning Source LLC
Chambersburg PA
CBHW031338230926
51848CB00002B/290
9788869775260